Advances in Anatomy, Embryology and Cell Biology
Ergebnisse der Anatomie und Entwicklungsgeschichte
Revues d'anatomie et de morphologie expérimentale

Springer-Verlag Berlin · Heidelberg · New York

This journal publishes reviews and critical articles covering the entire field of normal anatomy (cytology, histology, cyto- and histochemistry, electron microscopy, macroscopy, experimental morphology and embryology and comparative anatomy). Papers dealing with anthropology and clinical morphology will also be accepted with the aim of encouraging co-operation between anatomy and related disciplines.

Papers, which may be in English, French or German, are normally commissioned, but original papers and communications may be submitted and will be considered so long as they deal with a subject comprehensively and meet the requirements of the "Advances".

For speed of publication and breadth of distribution, this journal appears in single issues which can be purchased separately; 6 issues constitute one volume.

It is a fundamental condition that submitted manuscripts have not been, and will not simultaneously be submitted or published elsewhere. With the acceptance of a manuscript for publication, the publisher acquire full and exclusive copyright for all languages and countries.

25 copies of each paper are supplied free of charge.

Die Ergebnisse dienen der Veröffentlichung zusammenfassender und kritischer Artikel aus dem Gesamtgebiet der normalen Anatomie (Cytologie, Histologie, Cyto- und Histochemie, Elektronenmikroskopie, Makroskopie, experimentelle Morphologie und Embryologie und vergleichende Anatomie). Aufgenommen werden ferner Arbeiten anthropologischen und morphologisch-klinischen Inhalts, mit dem Ziel, die Zusammenarbeit zwischen Anatomie und Nachbardisziplinen zu fördern.

Zur Veröffentlichung gelangen in erster Linie angeforderte Manuskripte, jedoch werden auch eingesandte Arbeiten und Originalmitteilungen berücksichtigt, sofern sie ein Gebiet umfassend abhandeln und den Anforderungen der „Ergebnisse" genügen. Die Veröffentlichungen erfolgen in englischer, deutscher und französischer Sprache.

Die Arbeiten erscheinen im Interesse einer raschen Veröffentlichung und einer weiten Verbreitung als einzeln berechnete Hefte; je 6 Hefte bilden einen Band.

Grundsätzlich dürfen nur Arbeiten eingesandt werden, die nicht gleichzeitig an anderer Stelle zur Veröffentlichung eingereicht oder bereits veröffentlicht worden sind. Der Autor verpflichtet sich, seinen Beitrag auch nachträglich nicht an anderer Stelle zu publizieren.

Die Mitarbeiter erhalten von ihren Arbeiten zusammen 25 Freiexemplare.

Les résultats publient des sommaires et des articles critiques concernant l'ensemble du domaine de l'anatomie normale (cytologie, histologie, cyto- et histochimie, microscopie électronique, macroscopie, morphologie expérimentale, embryologie et anatomie comparée). Seront publiés en outre les articles traitant de l'anthropologie et de la morphologie clinique, en vue d'encourager la collaboration entre l'anatomie et les disciplines voisines.

Seront publiés en priorité les articles expressément demandés, nous tiendrons toutefois compte des articles qui nous seront envoyés dans la mesure où ils traitent d'un sejet dans son ensemble et correspondent aux standards des «Revues». Les publications seront faites en langues anglaise, allemande et française.

Dans l'intérêt d'une publication rapide et d'une large diffusion les travaux publiés paraitront dans des cahiers individuels, diffusés séparément: 6 cahiers forment un volume.

En principe, seuls les manuscrits qui n'ont encore été publiés ni dans le pays d'origine ni à l'éntranger peuvent nous être soumis. L'auteur s'engage en outre à ne pas les publier ailleurs ultérieument.

Les auteurs recevront 25 exemplaires gratuits de leur publication.

Manuscripts should be addressed to/Manuskripte sind zu senden an/Envoyer les manuscrits à:

Prof. Dr. A. BRODAL, Universitetet i Oslo, Anatomisk Institutt, Karl Johans Gate 47 (Domus Media), Oslo 1/Norwegen

Prof. W. HILD, Department of Anatomy, Medical Branch, The University of Texas, Galveston, Texas 77550/USA

Prof. Dr. J. van LIMBORGH, Universiteit van Amsterdam, Anatomisch-Embryologisch Laboratorium, Mauritskade 61, Amsterdam-O/Holland

Prof. Dr. R. ORTMANN, Anatomisches Institut der Universität, Lindenburg, D-5000 Köln-Lindenthal

Prof. Dr. T. H. SCHIEBLER, Anatomisches Institut der Universität, Koellikerstraße 6, D-8700 Würzburg

Prof. Dr. G. TÖNDURY, Direktion der Anatomie, Gloriastraße 19, CH-8006 Zürich/Schweiz

Advances in Anatomy, Embryology and Cell Biology
Ergebnisse der Anatomie und Entwicklungsgeschichte
Revues d'anatomie et de morphologie expérimentale

52/4

Editors: A. Brodal, Oslo · W. Hild, Galveston
J. van Limborgh, Amsterdam · R. Ortmann, Köln
T.H. Schiebler, Würzburg · G. Töndury, Zürich · E. Wolff, Paris

Finn-Mogens Šmejda Haug

Sulphide Silver Pattern and Cytoarchitectonics of Parahippocampal Areas in the Rat

Special Reference to the Subdivision of Area Entorhinalis
(Area 28) and its Demarcation from the Pyriform Cortex

With 49 Figures

Springer-Verlag Berlin Heidelberg GmbH 1976

Dr. Finn-Mogens Šmejda Haug, Associate Professor
Anatomical Institute
University of Oslo
Karl Johansgt. 47, Oslo 1, Norway

It is a pleasure to acknowledge the partial support of this study by USPHS Research grant NS 07998 and by the technical staffs of the anatomical institutes of Oslo and Aarhus.
This study has been referred to earlier as: Haug, F-M.Š.: Light microscopical mapping of the hippocampal region, the pyriform cortex and the corticomedial amygdaloid nuclei of the rat with Timm's sulphide silver method. II. Retrohippocampal areas.

ISBN 978-3-540-07850-0   ISBN 978-3-642-66448-9 (eBook)
DOI 10.1007/978-3-642-66448-9

Library of Congress Cataloging in Publication Data. Haug, Finn-Mogens Šmejda, 1937- Sulphide silver pattern and cytoarchitectonics of parahippocampal areas in the rat. (Advances in anatomy, embryology, and cell biology; v. 52, fasc. 4) Bibliography: p.   Includes index. 1. Hippocampus (Brain) 2. Stains and staining (Microscopy) 3. Rats-Anatomy. 4. Mammals-Cytology. I. Title. II. Title: Parahippocampal areas in the rat... III. Series. [DNLM: Limbic system—Anatomy and histology. 2. Stains and staining. W1 AD433K v. 52 fasc. 4 WL307 H368s] QL801.E67 vol. 52, fasc. 4 [QL933] 574.4'08s [599'.3233] 76-26676

Contents

Abbreviations and Symbols for all Figures

CA	anterior commissure	nlotp	cell aggregation along ventromedial corner of the hemisphere, just posterior to nlot (nlotp is called bed nucleus of the accessory olfactory tract by Scalia and Winans, 1975)
CA1, CA3	hippocampal fields according to Lorente de Nó (1934)		
CC	corpus callosum		
Co am	cortical nucleus of the amygdala		
F	fornix		
FI	fimbria	ob	olfactory bulb
LOT	lateral olfactory tract	ot	olfactory tubercle
OC	optic chiasm	p	posterior direction within the picture
OCC	occipital cortex		
OT	optic tract	pyr	pyriform cortex (see Nomenclature)
Psd	dorsal psalterium		
Psv	ventral psalterium	prsb	presubiculum
Spl	splenium of corpus callosum	s	septal area
ST	stria terminalis	sb	subiculum
TR	area interposed between area entorhinalis caudally and the pyriform cortex and the posterior pole of the amygdala rostrally	so	supraoptic nucleus of hypothalamus
		ts	triangular nucleus of the septum
		tt	tenia tecta (indusium griseum) **and** anterior continuation of the hippocampus (hippocampal rudiment)
aaa	anterior amygdaloid area		
am	medial nucleus of the amygdala	u	medial limit of area 28M' as characterized by cessation of the strong sulphide silver staining of layers III, II and inner zone of I
amf	amygdaloid fissure		
ci	the cingulum		
cl	claustrum		
d	1) dorsal direction within the picture 2) lamina dissecans (see Text) in the pre- and parasubiculum	x	1) bottom of posterior contiuation of the amygdaloid fissure 2) border in the molecular layer between the presubiculum and the subiculum
db	nucleus of the horizontal limb of the diagonal band of Broca		
e	lamina externa in the pre- and parasubiculum	y	border of area (rostroventrally in area entorhinalis) with patchy sulphide silver staining in layers II–IV
fd	fascia dentata		
hy	hypothalamus, a small part surfacing between horizontal limb of the diagonal band and the optic chiasm/tract, and including the supraoptic nucleus (so)	z	caudodorsal border of large area with a subpial white zone (area includes the pyriform cortex, the corticomedial amygdaloid nuclei and rostroventral part of area entorhinalis)
i	lamina interna in the pre- and parasubiculum	?	cortical area, dorsal to area 29e and the parasubiculum (possibly a ventral extension of 29c)
l	lateral direction within the picture		
m	medial direction within the picture		
neo	neocortex		
nlot	nucleus of the lateral olfactory tract	I–VI	layers of parahippocampal and (I–III) of pyriform cortex

7

Ia	glial cell rich superficial sublayer (thionine material) of pyriform cortex, recognizable also ventro-laterally in area entorhinalis	1–5	parts of cortical and medial nuclei of the amygdala (to be described in a separate article)
Ib	remaining deep sublayer of layer I (see Ia)	2–2'	lateral, respectively medial, alter-native for placing medial border of the dark stain in layer II of area 28L. The intervening transi-tional zone is stippled in the re-constructions (Figs. 1–10)
Id	deep, darkly sulphide silver stain-ed, subzone of layer I in pyriform and ventrolateral entorhinal cortex	28M, 28L	previously well established medial and lateral subdivisions of area entorhinalis (Blackstad, 1956)
Is	superficial, weakly sulphide silver stained, subzone of layer I (see Id). IsM and IsL in areas 28M and 28L	28M', 28L'	ventral subdivisions of area entor-hinalis identified in present report
IId	deep part of split layer II (thionine material) of area 28L	29b, 29c	retrosplenial (e.g., Krieg 1946) or cingulate (Domesick, 1969) cortex
IIs	superficial part of layer II (see IId)	29e	area retrosplenialis e (Blackstad,
IIIa	sublayer of layer III (thionine ma-terial) according to Lorente de Nó (1934)		1956, and others)
		49, 49a, 49b	parasubiculum and its subdivisions (Blackstad, 1956)
Vdl	dorsolateral border of layer V of area 28M	51e, 51f	(Krieg, 1946)

Symbols Used to Plot Interareal Borders in Parahippocampal Cortex and adjacent Pyriform Cortex

circles, filled	lateral limit of layer IV in areas 28M and 28M' (sulphide silver pattern)	diamonds, open	cytoarchitectonic limits of, or within, the pyriform cortex
circles, open	cytoarchitectonic limit of transi-tion area, TR	triangles, upright	cytoarchitectonic limits (in layer II) of area 28L and of the transi-tion between areas 28M and 28L.
diamonds, filled	caudodorsal limit of area in which the whole of layer III is darkly stained	triangles, inverted	cytoarchitectonic limit (in layer II) between areas 28M' and 28L'

I. Introduction

This study has two related objectives. One is to improve our understanding of the subdivision of the parahippocampal cortex, the other is to investigate the terminal distribution of sulphide silver stainable fibre systems (explained below) in this region.

The *parahippocampal areas* (comprising area entorhinalis, parasubiculum, area retrosplenialis e and presubiculum) transmit information to and from the hippocampus, a part of the brain which has been the subject of extensive neurobiological research. Much current anatomical work is therefore devoted to the study of the connections of the parahippocampal cortex (see Discussion), an activity which both requires and provides more precise concepts of its subdivision.

Recent studies have shown that histochemistry often brings out laminae and areas in this cortical region more clearly than do conventional morphological methods (Storm-Mathisen and Blackstad, 1964; Mellgren and Blackstad, 1967; Geneser Jensen and Blackstad, 1971; Geneser Jensen et al., 1974 and references therein; Mellgren, 1973 a, b). The sulphide silver method, used here, is particularly valuable in this respect, as will be explained shortly.

When using histochemical techniques in this way, it is essential to correlate the staining pattern with subdivisions obtained—or obtainable—by conventional morphological methods. Ideally, several histochemical and morphological methods, demonstrating different aspects of the tissue structure, should be applied to identical or adjacent sections of the same brain to permit a satisfactory comparison of the various staining patterns. In practice, this requirement can hardly be met. Thus, the present paper compares the cytoarchitectonic and sulphide silver patterns in adjacent sections of the same brains.

Despite much confusion resulting from attempts to subdivide the hippocampal region solely on the basis of cytoarchitectonics, *Nissl staining* does provide important information on cellular architecture particularly when combined with other methods. In the present study systematic comparison of the cyto- and chemoarchitectonic borders is aided by construction of *surface maps* based on serial sections of the hippocampal and adjacent cortical regions. The resulting diagrams also facilitate comparisons with other authors' subdivisions of the parahippocampal cortex and are therefore an important part of the observations to be presented here.

The sulphide silver method, developed to visualize normal and toxic metal-distributions in tissue sections (Timm, 1958a, b; Voigt, 1951), stains neuropil in most parts of the forebrain (Haug, 1973). The staining pattern shows conspicuous laminar and regional differentiation, suggesting an association of the neuropil stain with fibre systems that terminate in the stained laminae and areas. The correlation between laminar staining and laminar synaptic fields is particularly suggestive in the hippocampus and fascia dentata (Haug, 1973, 1974). Moreover, in the latter areas experimental anterograde degeneration of various fibre systems causes rapid loss of neuropil stain in the denervated laminae (Haug, 1975, 1976 and in preparation). This confirms that the stain depends on the transected fibre systems and suggests, but does not prove, a presynaptic localization of the stain. I shall here use the term "sulphide silver positive fibre systems" to denote fibre systems that cause their terminal fields to be stainable with Timm's method, irrespective of the exact relation of the stain to the synapses. It should be added that both the finer light microscopic distribution of the stain throughout the forebrain and preliminary electron microscopic findings (restricted to stratum radiatum of the hippocampus) suggest a presynaptic localization (Haug, 1975, 1976).

The above evidence points to the conclusion that describing laminar and regional distribution of sulphide silver staining in the hippocampal formation and other forebrain areas is equivalent to describing the form and density of selected synaptic fields. Thus, we may assume that the borders and transitions to be reported here reflect borders and transitions in synaptic architecture and therefore are functionally important.

Strongly related to the primary purpose of the present paper is its secondary purpose, which is to provide an extensive morphological basis for the investigation of the functional correlates of the sulphide silver staining in the brain. Ultrastructural localization of the stain and experimental identification of the origin and course of the sulphide silver positive fibre systems will require exact maps of the normal sulphide silver pattern, as provided here for the parahippocampal areas.

II. Material and Methods

A. Animals. Histological Methods

Adult albino rats (Wistar) of both sexes were anaesthetized with ether and Nembutal and perfused transcardially with a phosphate buffered sulphide solution. 40 μm thick cryostat sections of the frozen brains were all serially mounted on microscope slides and silver impregnated [1] by physical development. A detailed description of the modified Timm's (1958 a) sulphide silver method was given elsewhere (Haug, 1973). As a rule, three parallel series were formed from each brain. Two of the three series were sulphide silver stained, respectively "lightly" and "darkly", and one series stained with thionin. Thus, of three consecutive sections two sections were sulphide silver stained and one stained with thionin.

The present observations are based on 9 horizontally [2], 6 frontally and 4 parasagittally sectioned brains. Three of these brains (employed also in a preceeding paper, Haug, 1974) were used for all of the present photomicrographs and for the graphic reconstructions. Several other brains have been examined for other purposes in the course of experimental studies on the sulphide silver pattern in the hippocampal region.

---

[1] In the following the term "stain" is preferred to "impregnate, although the latter is strictly more correct.

[2] Fig. 9 shows what is here meant by "horizontal". This plane, which differs somewhat from the horizontal plane of König and Klippel (1963), resulted from placing the soft brain with the basal surface on the specimen holder of the cryostat.

B. Graphic Reconstructions

The right hemispheres of three brains, sectioned horizontally, frontally and parasagittally, were reconstructed graphically as follows. For each brain, line drawings or photomicrographs were made of all the sections at a magnification of 25 times. These reproductions were placed sequentially on a drawing board and points projected from them on to sets of equidistant, straight, lines that represented the desired projections. In this way, the lateral and medial (brainstem removed) surfaces of a hemisphere were projected on to a parasagittal plane, the posterior surface on to a frontal plane, the ventral surface on to a horizontal plane or the posteroventral surface on to a plane intermediate between frontal and horizontal. Figs. 3–10 show which projections were constructed for each of the brains. To simplify the plots, each thionine stained section with its two adjacent Timm stained neighbours (all 40 µm thick) was considered as one *level* and represented as a single line in the projections. The equidistance in the original reconstructions thus became 120 µm x the magnification, i. e., approximately 3 mm. To ensure the correct relative orientation of the reproductions within a series, reference holes had been drilled through the frozen brains before they were sectioned.

Figs. 1 and 2 show how the reconstructions turned out. Minor distortions of the sections and small errors in superposing them caused all curves to become slightly serrated. Some curves had larger irregularities, particularly the tentative limits of transition zones difficult to define. In the final diagrams (Figs. 3–10) minor irregularities were completely smoothed out. To make the drawings legible, the quite irregular cytoarchitectonic borders of the transition between areas 28M and 28L (triangles, Figs. 1–2, 4 and 9) were also completely straightened out.

It is obviously desirable to asess the variability (e. g., between observers, between brains) of the borders. However, the present procedures are too slow for such a larger scale study and for many other routine uses of the reconstruction approach. Computer graphics should be used for this, perhaps one of the current systems for computer aided anatomical reconstruction (Katz and Levinthal, 1972; Llinás and Hillman, 1975; Rakic et al., 1974).

Sulphide perfused brains are soft, as if unfixed, and the brains used here were therefore somewhat flattened by their own weight when placed on the specimen holder prior to freezing. Aldehyde fixation of the brains was avoided in the present study since it slightly weakens the sulphide silver stainability. However, a brief aldehyde fixation may be used if it is more important to perserve the form of the brains than the full intensity of sulphide silver stainability.

Despite the sources of error just mentioned, there is good qualitative agreement between comparable projections based on the three different brains.

C. Nomenclature

In the present study the term hippocampal *formation* denotes the fascia dentata, hippocampus and the subiculum; the term hippocampal *region* includes in addition the presubiculum, area 29e, the parasubiculum and the entorhinal area (Blackstad, 1956; Shipley, 1975); the term parahippocampal is synonymous with retrohippocampal. The terminology adopted by Blackstad and co-workers (Blackstad, 1956; Storm-Mathisen and Blackstad, 1964; Mellgren and Blackstad, 1967; Geneser Jensen and Blackstad, 1971; Haug, 1974) to designate the layers and areas in the hippocampal region is followed in the present paper, if not otherwise indicated. A revision of this terminology has been proposed by Stephan (1975). The important term *lamina dissecans* needs explicit definition. It is taken to comprise both laminae IIIa und IV wherever these layers are separable. This roughly corresponds to the way M. Rose (1929, Tafel 14–17; 1931, several figures) identified lamina dissecans in the mature entorhinal area, although on some illustrations he also included the present layer III in it (M. Rose, 1931, Tafel 27, 30). Stephan (1975, pp. 202–206, 650–653, 661, 665, 694) succinctly discussed M. Rose's and subsequent investigators' descriptions of the ontogenesis and comparative anatomy of lamina dissecans and illustrated the cytoarchitectonic appearance of this layer in various parts of the parahippocampal cortex of selected species.

The terms pyriform cortex and prepyriform cortex are here used synonymously and in the same sense as Pigache (1970) used "primary olfactory cortex", Price and Powell (1971) "pyriform cortex", Price (1973) "prepyriform cortex" and Stephan (1975, pp. 428–434, 441–442) "regio praepiriformis". In this terminology there is no periamygdaloid cortex lateral to the fissura amygdala (sulcus semiannularis). Consult Pigache (1970) and Stephan (1975) for thorough reviews and critical discussions of the many terminologies previously applied to the cortex of the pyriform lobe.

III. Observations

In *thionine stained* sections prepared from unfixed and frozen brains there is considerable loss of cytological detail. The "cytoarchitectonic" characteristics observable in the present material are, therefore, mostly cell-packing and to some extent cell-size. Subdivision on these criteria are often best performed in rather thick sections. Trials with sections thicker than 40 µm did not, however, disclose new features.

The *sulphide silver staining* in the rat forebrain *neuropil* appears as grains ranging in size up to about 1 µm (rarely larger as in the hippocampal mossy fibre layer). Stained granules are also located in *neuronal somata,* but usually without contributing much to the overall staining density of the cortical laminae. There are *regional* and *laminar* differences in the staining of somata. Moreover, although the principal features of the staining patterns are constant, some series show a higher ratio between the staining intensities of somata in general and neuropil than those shown here. The reasons for this variation are unknown; both biological and methodological factors may contribute (Haug, 1973). Within the rat forebrain, *glial cell* bodies and processes contribute little to the staining pattern. For further details on the fine distribution of the staining product, see Haug (1973).

The present account is mainly concerned with the granular staining of the neuropil, which in all probability represents stained synaptic boutons or spines, as explained above. Staining of neuronal somata will be noted in passing for some laminae.

A. Parahippocampal Areas except Area Entorhinalis

Under this heading will be described the presubiculum, area retrosplenialis e and the parasubiculum.[3]

1. Cytoarchitectonics (Figs. 11f–23b, 29b–34b, 44b–48j, k, l)[4]

The present material largely confirms previous observations on the cytoarchitecture of the three areas. A brief review of their *laminar subdivision* will facilitate the subsequent description of the Timm-material.

[3] Stephan (1975, pp. 716 ff.) and others designated these areas together as Regio praesubicularis.

[4] *General explanation for all photomicrographs (Figs. 11–48).* Figs. 11–28 illustrates a horizontal series, Figs. 29–39 a frontal series and Figs. 40–48 a sagittal series through the hippocampal region and immediately adjacent structures. Pairs of adjacent Timm and thionine stained cryostat sec-

12

Layer I is the superficial plexiform, or molecular, layer seen in all three areas at all dorsoventral levels.

A narrow *layer II* may be distinguished from the underlying broad layer III only in the presubiculum (e. g. II, Figs. 12b—17b). It shows clusters (e. g., flanking arrows Figs. 16b, 18b), particularly near the subiculum.

In all three areas *layer III* is broad. A narrow *deep plexiform layer* (IIIa, Fig. 14b) is seen more constantly in the parasubiculum than in the presubiculum and this holds both for dorsal and ventral levels (cp. Figs. 14b—16b with Figs. 18b—21b). It is continuous with layer IIIa of area 28M (see below).

A very narrow *layer IV* of somewhat larger cells than in the layers above or deep to it is seen dorsally in the parasubiculum, area retrosplenialis e and presubiculum (IV, 14b—15b), but not ventrally (Fig. 19b). This dorsoventral change parallels that to be described in areas 28M—28M'.

A combined *layer V/VI* is seen in all three areas at all dorsoventral levels, the designation V/VI being used here only for recalling the continuity with the deep entorhinal cortex in which laminae V and VI may be separated.

In the above description, Cajal's (1911, 1955) subdivision of the presubiculum into 5 laminae has been slightly modified, by using roman numerals and by defining a sublayer IIIa in analogy with that of area 28M. Evidently, this scheme best describes *dorsal levels* of the *presubiculum,* while ventrally and in areas 29e and 49 one or more of laminae II, IIIa or IV are ill defined or absent. (See Nomenclature concerning the definition of lamina dissecans.)

For *interareal limits* the reader is referred to the illustrations. *Area 29e* is in general difficult to define cytoarchitectonically. Nevertheless is may often be tentatively delimited from the presubiculum and the parasubiculum with borders which are confirmed by the sulphide silver pattern (see below).

The other interareal limits of the *pre- and parasubiculum* are mostly clearcut. However, at far ventral levels (Figs. 22b—23b, 33b—34b) where the laminar structures of both areas are simpler than dorsally there are no unequivocal limit between the pre- and parasubiculum or between the latter and the entorhinal area.

The three areas differ from each other only with regard to the lamina externa; laminae dissecans and interna show no cytoarchitectonic interareal borders. Certainly, at dorsal levels, one can delimit a sector of layer V/VI as belonging to each area, by drawing lines perpendicular to the pia. The resulting borders, however, are arbitrary, since they are not based on structural changes within the lamina interna.

---

tions, 40 μm thick, are shown. Small numbers in upper left or right hand corner give the level within the series, permitting its precise localization on the corresponding reconstructions. The Timm stained section precedes a thionine stained section of the same level number, except in the sagittal series, where *a* (Timm) preceeds *b* (thionine) which preceeds *c* (Timm). Figure numbers are given in bold characters in lower left or right hand corner. Symbols and abbreviations (see list) correspond to those used in the reconstructions. In those cases where interareal borders seemed vague, tentative judgements are often indicated by using broken rather than continuous lines. In some places the presence of transitional zones are indicated by labelling the least and the most conservative limit of the area to be defined. To facilitate the comparison between adjacent Timm- and Nissl-stained sections, border marks have often been transferred from one to the other (bordermarks labelled with asterisks). For the sulphide silver stained sections, a broken line often indicates the superficial limit of the molecular layer; it was plotted in the photographs on the basis of direct microscopic examination of the sections. Small parts of a picture marked with a cross or encircled with a broken line have been retouched because of distracting technical defects in the sections. In general, the micrographs have been orientated so that either the posterior or dorsal direction is upwards in the picture. Where this was not practicable, directions are indicated by orthogonal arrows.

2. Sulphide Silver Pattern

a) Presubiculum (Figs. 11e–22a, 30a–34a, 44a–48k)

In *layer I* the neuropil is sprinkled with distinct, fairly large, dark, grains and appears weakly greyish-brown at low magnifications. A sharp border is formed where the staining ceases towards the outer zone of the subiculum's molecular layer (x, Fig. 17a and elsewhere). This border corresponds to that found by Blackstad in normal Nauta stained material and for degenerating afferents to the subiculum (Fig. 7 of Blackstad, 1956). It requires fairly strong development to be visualized, but is then apparent at all levels (x, Figs. 12a–21a) except near the ventral end of the presubiculum where the stain in layer I is often reduced (Fig. 34a).

In *layer II* a slightly enhanced staining of the neuropil together with some staining of cell somata causes a darker hue than in both layers I and III (II, Figs. 13a, 19a, 30a, 31a). As defined by the sulphide silver stain, layer II appears slightly thicker than with the Nissl stain, especially at more ventral levels. The neuropil stain continues lateralwards into the even darker area 29e (see below) but, when followed medialwards, tends to cease somewhat lateral to the subicular border. It is noteworthy that where the layer is broken up into cellular islands, the neuropil stain is associated with these islands and not with the intervening cell-poor stretches (suggested by arrow Figs. 16a,b, 18a,b).

Layer III is clearly darker than layer I at low magnifications. This is probably mostly due to the rather strong staining of cell bodies in this layer. Nevertheless, higher magnifications suggest that even the neuropil is a little darker in layer III than in layer I. The limit towards the subiculum was described previously (Haug, 1974). The limit towards the parasubiculum is marked by the interposition of the triangular area 29e superficially (Figs. 14a–16a, see also 17a–18a).

Layer IV appears as a narrow zone of slightly enhanced staining at the bottom of layer III (d, Figs. 18a–20a, 31a–33a). Unlike in area 28M (see below) it is not separated from layer III by a pale stripe. Medially layer IV merges with the inner zone of the subiculum (Haug, 1974, p. 21, last paragraph). The neuropil stain of layer IV is often somewhat stronger than in the present illustrations and visible even at dorsal levels where it appears to be absent from the present horizontal sections (Figs. 12a–16a).

Somata are also stained but this is more evident in series which show a higher ratio between the staining intensities of somata and of neuropil than do those shown here. In such series, somata in the adjoining part of subiculum's inner zone have the same rather strong dotted staining as those of presubiculum's layer IV so that these two layers fuse both with respect to the staining of neuropil and of somata.

Layer V/VI is continuous with the corresponding layer of the parasubiculum. It tapers towards the subiculum and extends for a short distance into the latter as previously described (Haug, 1974 loc. cit.). No distinction between a Vth and a VIth layer is possible, although the stain is often somewhat reduced near the white matter.

The dorsoventral extent of the area is well brought out by the Timm pattern (Figs. 5, 8, 10 and the corresponding photomicrographs). As in the cytoarchitectonic pattern there are no principal dorsoventral changes, only a gradual reduction in the overall thickness of the cortex and the thickness of each layer. Layer II tends to become broader and darker at ventral levels. The area is buried in the hippocampal fissure and reduced in width to a small wedge of uncharacteristic lamination shortly before it is lost between the subiculum and the parasubiculum.

b) Area 29e (Figs. 11e–16a, 29a–31a, 46a–48g)

This is a triangular part of the lamina externa with its base superficially and its apex interposed between the presubicular and the parasubicular layer III. It is difficult to delimit in most Niss-stained sections (Figs. 12b–16b, 29b–31b), but has been well characterized with other methods (see discussion).

In sulphide silver preparations (Figs. 14a–16a, 29a–31a) lamina externa of area 29e is dark. The dark stain grades into layer II of the presubiculum (Figs. 16a, 30a, 31a) and shows an edge-sharp border towards the parasubiculum. The neuropil is particularly densely stained, thus masking the staining of neuronal somata. This stain also encroaches on the deeper levels of layer I (cp. Figs. 14a and b, 15a and b), but thence fades quickly towards the surface.

Dorsally, area 29e borders the occipital neocortex (OCC, Figs. 4, 5, 10 is area 18a according to Figs. 1, 4, 5 of Krieg, 1946) and barely touches another area, which may possibly be a posteroventral extension of area 29c (questionmark in Figs. 4, 5, 10 and 11). Area 29b is in contact with the presubiculum rather than with area 29e. The very dorsal end of area 29e can not be delimited with certainty against the equally darkly stained occipital cortex (Figs. 11e–13a). Only tentative limits have been indicated.

When area 29e is traced ventrally (Fig. 16a) it remains as a triangular patch of dark sulphide silver staining farther than one might expect from a study of normal fibre architecture (Blackstad, 1956, Fig. 6B).

c) Parasubiculum (Figs. 14a–23a, 29a–35a, 46a–48c)

Lamina externa differs between subareas a and b:
In *Parasubiculum b* this lamina stains intermediately (Figs. 16a–18a), neuropil contributing more than somata to the overall density. The superficial half of the molecular layer (cp. Figs. 16a and b; or 18a and b) remains weakly stained as in adjoining areas. The medial border of 49b was described above. In all brains there is an intensification of the neuropil stain in a small area near the medial border at some ventral levels (Figs. 18a–21a).

In *parasubiculum a* the staining intensity of lamina externa increases from near the lamina dissecans to reach a maximun within the deep third of layer I (Figs. 18a,b, 19a, b). Superficial to this, the intensity is abruptly reduced but often still remains slightly darker than lateral and medial to area 49a. Layer I as a whole and its two subzones are narrower in area 49a than in the adjacent areas (Blackstad, 1956).

The border of area 49a is usually sharp towards the entorhinal area and blurred towards the parasubiculum b. If parasubiculum a is defined as the darkly sulphide silver stained triangle it tapers to finally disappear at dorsalmost levels (Figs. 17a–14a) or, conversely, widens towards its ventral end (Figs. 17a–21a). At the ventralmost levels the staining increases somewhat near the presubiculum and decreases near the entorhinal area, reversing the relative densities of areas 49a and b. The somewhat vague external and internal limits of area 49a at these levels are tentatively indicated in Figs. 22a–23a and 34a–35a.

Lamina dissecans may best be visualized as a layer at more dorsal levels (Figs. 14b–16b). It tends to be slightly paler than the lamina externa and is continuous with lamina dissecans of area 28 M (see below).

A thin, pale stripe extends superficially from the lamina dissecans between the parasubiculum and area 28M, contributing to the sharp delimitation of these two areas from each other. Nissl sections (e. g. the present figures) and previous histochemical studies (Mellgren and Blackstad, 1967) show the same pale stripe.

In *lamina interna* there are no recognizable limits of the parasubiculum or between its subfields a and b. Rather, the lamina interna appears to be continuous from the entorhinal area through the parasubiculum and into the presubiculum, showing a gradual decrease of sulphide silver staining in the medial direction.

B. Area Entorhinalis and its Transition to the Pyriform Cortex and the Amygdala

1. Cytoarchitectonics

In recent years the subdivision of area 28 into medial and lateral parts has received support from several lines of study, although the presence of a transitional zone has been suggested (see Discussion).

The present observations support this subdivision, and the suggestions of an intermediate zone, as regards the dorsal, major, parts of area entorhinalis. Rostroventrally, however, two further subdivisions (areas 28M' and 28L') and an area (TR) interposed between the entorhinal and pyriform cortices will be defined.

a) Medial Part, Area 28M (Figs. 13b–23b, 29b–32b, 41b–47b)

In this area Cajal (1911, 1955) described seven layers. These were renumbered into six by Lorente de Nó (1933), and since repeatedly confirmed morphologically (Blackstad, 1956) and histochemically (Storm-Mathisen and Blackstad, 1964; Mellgren and Blackstad, 1967; Geneser-Jensen et al., 1974).

A few selected features of this well known laminar cytoarchitecture (Cajal, 1911, 1955; Lorente de Nó, 1933) will be emphasized to facilitate the subsequent descriptions of cytoarchitectonic borders and transitions to other areas and of the sulphide silver pattern.

A compact, *layer II* (Fig. 16b) is prominent in most of area 28M. It becomes gradually less distinct in the medioventral part of this area (Fig. 21b).

Layer III is broad and homogeneous and characteristically well delimited both superficially and deeply. Its deep border is towards the cell poor sublayer IIIa.

Sublayer IIIa is Cajal's (1911, 1955) layer 4 or *deep plexiform layer*. It is a very discrete light stripe almost devoid of neuronal somata, and is one of the best characteristics of area 28M in Nissl material. Its present designation, layer IIIa, was chosen by Lorente de Nó (1933) to emphasize its close similarity to the main layer III: It contains dendrites of the more basally situated pyramidal cells in layer III and shares the axonal plexus of the latter layer.

Layer IV (5 of Cajal, 1911, 1955) is several times thicker than sublayer IIIa—except medially—and contains scattered cells, larger than in the adjacent layers.

At its lateral end, layer IV is five or more cell diameters deep. Medialwards it shrinks gradually to be hardly distinguishable from the overlying layer IIIa in the vicinity of the parasubiculum.

While the border of layer IV towards layer IIIa is conspicuous and sharp, that towards layer V is often less obvious at first sight, since there is no cell poor stripe here. Nevertheless, closer study usually reveals a fairly sharp border (Figs. 17b, 19b, 30b, 31), especially on using higher magnification to bring out the change in cell size.

Layer V (6 of Cajal, 1911, 1955) shows numerous small, tightly packed, cells.

Layer VI (7 of Cajal, 1911, 1955) is barely distinguishable from layer V due to a slightly looser packing of its somata (VI, Figs. 18b–21b, 31b).

The limits of area 28 M are as follows: The *medial limit* (Figs. 5, 8 and corresponding photos) is fairly sharp within lamina externa, except far ventrally (Fig. 21b). As will be seen below, the sulphide silver pattern reveals an even sharper border.

The *dorsal and dorsolateral* limits (Fig. 49, 4, 9 and corresponding photos) are also cytoarchitectonically conspicuous within layers II/III and V. The sharp, laterally convex, limit of layers II and III is continuous with the deep limit of layer III (arrow Figs. 16b–17b), a superficial extension of layer IIIa being interposed between the entorhinal area and the neocortex. Only at the dorsalmost levels is the lateral border of layers II and III somewhat uncertain (Figs. 15b–13b, 43b–44b). The dorsal and lateral limits of layer IV are not sharp in the present material (Figs. 14b–17b, 42b–44b). Layer V is sharply delimited dorsolaterally as seen in sagittal sections (Vdl, Figs. 42b–44b) and less clearly in frontal (Figs. 29b–31b) and horizontal (Figs. 14b–17b) sections (compare also with the adjacent sulphide silver stained sections).

The lateral, ventrolateral, and ventromedial limits of area 28M are described below.

b) Lateral Part, Area 28L (Figs. 18b–24b, 30b–35d, 41b–45b)

The most typical features of this part, as compared to area 28M, are the following (Figs. 21b–22b, 31b–32b).

Layer II is incompletely split into two sublayers. The outer sublayer (labelled IIs) tends to break up into cellular groups or islands, the inner one (labelled IId) is really not sharply distinguishable from layer III (labelled III).

Layers IIIa (deep plexiform layer of Cajal, 1955) *and IV* are less evident in the dorsomedial half of area 28L (Figs. 19b–21b, 31b) than in area 28M, and become even more vaguely defined further ventrally (Figs. 21b–23b, 32b–33b, 41b, 42b).

Layers V and VI may be differentiated near 28M (Figs. 19b–21b) but not further ventrally (Figs. 21b–23b).

The *limits of area 28 L* are (Figs. 49, 3, 4, 6, 7, 9, 10 and corresponding photomicrographs): *Dorsomedially* towards area 28 M, *dorsolaterally* towards the area perirhinalis (see p. 30) along the rhinal fissure, *ventrolaterally* towards the pyriform cortex, *ventrally* (rostrally) towards area 28L' and *ventromedially* towards area 28M' as defined below. Of these borders, only the *dorsolateral* one is sharp and clear. Here the outer sublayer of layer II ceases abruptly near the rhinal fissure, the cells of the deep sublayer of II and of layer III are suddenly replaced by smaller and more densely packed cells and even layer IV has a recognizable lateral limit (Fig. 21b). The lamina interna fails to show a clear border.

The other borders mentioned are described below.

c) Transition between Medial and Lateral Parts of Area 28
(Figs. 18b–22b, 29b–32b, 41b–44b)

Although the medial and lateral parts are easily distinguished, there is no sharp limit between them in the present Nissl stained material. On the photographs two tentative limits between areas 28L and 28M are therefore indicated in layer II, one more conservative (restrictive) than the other with respect to area 28L (upright triangles, Figs. 19b–23b, 30b–33b, 41b–45b). In the reconstructions (Figs. 3, 4, 6, 7, 9, 10) this reappears as a "transitional zone" (enclosed by upright triangles) difficult to assign with certainty to either areas 28M or 28L on the basis of layer II–cytoarchitectonics. The appearance of the lateral border of lamina externa also changes gradually from area 28M to area 28L. In the former this border is laterally convex and especially prominent because of the superficial extension of layer IIIa, in the latter the border is straight and less conspicuous because of the absence of the superficial extension of layer IIIa (Figs. 16b–21b).

d) Definition of a New Ventromedial Part, Area 28M'
(Figs. 24b–27b, 33b–36b, 45b–47b)

There is a striking cytoarchitectonic difference between typical area 28M and the ventromedial corner of the entorhinal area (Figs. 4, 7, 9, 10): Ventromedially (Figs. 24b–27b) the cortex is thinner than dorsally. It lacks a distinct layer II within the narrow lamina externa (II/III, Fig. 25), has a narrow, fairly well defined lamina dissecans (combined layers IIIa/IV; labelled IIIa in Figs. 24b–26b) and a narrow lamina interna which is hardly divisible into layers V and VI (Figs. 25b–27b) except dorsally (V, VI, Fig. 24b).

Serial sections in the three standard planes of sectioning show a transition rather than an unequivocal border between the classical area 28M and the ventromedial part which will here be called area 28M'. There is actually a cytoarchitectonic gradient over an extensive dorsoventral range, particularly with regard to layer II (Figs. 15b–23b, 45b–46b). At certain ventral levels, however, a particularly rapid change in stratification occurs permitting a fairly clear separation of area 28M from area 28M' (between Figs. 21b–23b).

The structure of area 28M' is not unlike that of the parasubiculum from which it may be difficult to separate by cytoarchitectonic criteria alone (Figs. 22b–23b, 34b). However, in the sulphide silver pattern these two areas are clearly different (Figs. 22a–24a, 34a–35a). Furthermore in its laminar cytoarchitecture (and sulphide silver pattern) area 28M' is as different from area 28L as it is from area 28M. The designation 28M' was chosen for topographical and mnemotechnical reasons.

The limits of area 28M' are then (Figs. 49, 4, 5, 7, 8, 9, 10): *Dorsally* towards area 28M, *dorsomedially* towards the parasubiculum, *ventromedially* towards the subiculum *rostrally* towards the posterior pole of the cortical amygdaloid nucleus and *laterally* towards areas 28L, 28L' and the transition area TR (see below), at successively more rostroventral levels. These borders are not sharp, except in lamina externa towards the subiculum and perhaps in the lateral end of layer IV towards the area TR.

e) "Transitional" Area, TR, between Areas 28, 51 and the Amygdala
(Figs. 25b–28b, 36b–38b, 42b–44b)

This cortical area forms the bottom of the wide posterior extension of the amygdaloid fissure. It is thinner than the cortical areas caudally and laterally to it and appears to consist only of the molecular layer and a cell layer which may be traced into the lamina externa of the surrounding cortex (see description of the transition below).

Rostrally (Figs. 37b–38b) the cell layer of area TR extends medially deep to the posterior part of the cortical nucleus (TR, Fig. 38). It also becomes continuous with a somewhat irregular sector of the pyriform cortex which lies deep to the amygdaloid fissure (forwards of amf, Fig. 39a,b) and has been termed the corticoamygdaloid transition area (e.g. T in Figs. 1, 4, 7, 10 of Hall and Geneser Jensen, 1971) or area semi-annularis, Pam Cs (Stephan, 1975, p. 357).

f) Definition of a New Ventrolateral Part, Area 28L' (Figs. 23b–25b, 34b–37b, 41b–44b)

It is apparent from the illustrations that the cytoarchitectonic structure typical of area 28L changes gradually into the well defined adjoining areas which are area 28M, the pyriform cortex, area TR and area 28M' (Figs. 49, 3, 4, 6, 7, 9, 10 and the photomacrographs). Here are considered the transitions into a) the pyriform cortex and b) area TR.

α) *Laterally* the following gradual changes take place as the cortex is followed from typical area 28L into typical pyriform cortex (Figs. 22b–26b, 32b–39b):

1) A reunification of the split layer II into a very compact, cell rich layer. 2) Layer IV finally becomes indiscernible. 3) Lamina interna (V–VI, Figs. 35d–38b) merges with a cell mass (cl, Figs. 26b–27b, 39b) which lies just superficial to the external capsule and is again covered with a cell poor layer. This cell mass corresponds to the endopyriform nucleus of Loo (1931, Figs. 33, 35, 38 and pp. 63–64) and is included in the claustrum of other authors (e.g., Stephan, 1975, Figs. 290, 270, 284, and pp. 435, 443, 464, 483–485). 4) The depth of the cortex is not reduced.

Between typical entorhinal and typical pyriform cortex lies a zone which can not be definitely assigned to either one of these two areas. Thus, proceeding rostralwards from the entorhinal area one may indentify layers II, III, IV and the lamina interna as far forwards as in Figs. 36b–38b. Only in Fig. 39b is the typical three-layered pyriform cortex well established. On the other hand, proceeding caudalwards from the pyriform cortex into the entorhinal area, one may choose to recognize the typical entorhinal layering only caudal to Fig. 36. The extent of the area which is difficult to assign with certainty to either area 28L or the pyriform cortex (between Figs. 36 and 39) has been indicated by stippling in Figs. 6 and 7.

β) *Medially in area 28L* the following changes take place as the area TR (Figs. 26b, 36b) is approached (Figs. 23b–25b, 33b–35d):

1) A gradual thinning out of the superficial sublayer of layer II (Figs. 33b–35d). 2) A looser cell packing in the deep sublayer (IId–III, Figs. 33b–34b) and 3) Subsequent reestablishment of layer II ("lamina externa") which gradually thickens and becomes more compact (II, Figs. 36b, 37b). 4) At the same time layer IV disappears and the combined layers V/VI are reduced to an insignificant thickness (Figs. 35d–37b). 5) The total depth of the cortex is reduced as a result of the above changes.

It is unsatisfactory to reduce continuous changes of architecture to discrete borders As a compromise, a rostroventral limit is here placed where area 28L ceases to be typical, excluding from it a narrow zone of transition into the pyriform cortex laterally and into the area TR medially. This transitional zone is tentatively labelled area 28L' (see Discussion).

2. Sulphide Silver Pattern

a) Medial Part, Area 28M (Figs. 13a–23a, 29a–32a, 41a–47a)

Layer I shows weak sulphide silver staining of its superficial half (Is, Fig. 16a) as in the parasubiculum and presubiculum. The staining density increases on entering the deep half of layer I (Id, Fig. 16a), roughly at the limit between the sublaminae supratangentialis and tangentialis as identified by earlier authors (see Blackstad, 1956). Like sublamina supratangentialis the outer weakly Timm-stained part of layer I tapers on approaching the parasubiculum. Laterally the superficial pale zone continues into the neocortex where, however, it becomes darker, narrower and less distinctly delimited from the inner dense part of layer I (e.g. Fig. 17a). Indeed, a quite sharp rise in staining density is often seen in the superficial zone of layer I at the limit between area entorhinalis and the neocortex (Figs. 16a–20a).

Layer II as a whole appears slightly lighter than the adjacent layers. From the present material it cannot be determined whether this is only a result of the lower volume fraction of neuropil in this cell rich layer or if the remaining neuropil in addition is less stainable than in the adjacent laminae. At its medial, broader, end layer II (IIm, Figs. 17a–18a), and even the inner half of I, becomes paler. This does seem to be due to a decrease in staining of both neuropil and somata. Somata in this layer show a dotted staining.

In layer III the neuropil stains more densely than in I–II. Furthermore, at somewhat ventral levels, there is a weak lateromedial gradient of *increasing* density (reverse of what is seen in layers I/II) which enhances the difference between layers III (IIIm) and I/II (IIm) medially (Figs. 15a–21a).

Most of the cell poor sublayer IIIa appears to be part of layer III, but a thin pale line extends along the deep and medial limits of the latter (a, Fig. 18a and elsewhere). The present material does not show whether this thin line belongs to Lorente de Nó's (1933) layer IIIa or his IV, because the border between layers IIIa and IV can not be drawn with the required precision in Nissl-stained sections. In the following, the white line is assigned to layer IV (see below). Laterally, layer III is sharply delimited where it borders the darkly staining neocortex (area perirhinalis, see below), but has no sharp border towards area 28L.

The majority of somata in layer III show a dotted staining, though less conspicuous than in layer II.

Layer IV show a subtle sublamination of the neuropil stain, most apparent laterally in area 28M: If this layer is defined as in Fig. 18a it includes a) the superficial and very discrete pale line mentioned above, b) a broader intervening zone of intermediate staining, c) a deep, somewhat blurred, pale line (a, b, c, Fig. 18a); the lines a and c are even better seen in paraffin-embedded material or tissue given a short aldehyde fixation prior to freezing (not shown here). Laterally, the middle zone (b) forms the main part of layer IV (arrow labelled IV in Fig. 18a; see also Fig. 17a). Further medially, as layer IV shrinks, the middle zone dwindles, usually allowing the two delimiting pale lines to fuse in the narrow medial part of layer IV (below IIm, Fig. 18a). Occasionally the middle zone may be traced as an extremely thin line into lamina dissecans of the parasubiculum (not shown). The superficial and deep pale lines may perhaps represent the thin white zones said to be free of dendritic plexuses and to consist almost only of glial cells (Lorente de Nó, 1933, p. 431, last sentence in text). In any case these lines are remarkable, having no analogue in any other cortical area.

At the lateral end of layer IV the stain in zone b increases rapidly to the dark stain that covers the adjoining neocortical layers II–IV (surrounding arrows, Figs. 16a–17a). Concurrently, the superficial limit of zone b bends towards the pia to continue smoothly in the sharp, medially concave, limit of the dark neocortical staining just mentioned. However, the pale line (a) usually disappears laterally in layer IV.

All the *cell somata* of layer IV are devoid of granular staining, thus contrasting with most somata of the other entorhinal layers.

Layer V. The superficial limit of layer V is unequivocal as just described. Within the layer there is a gradual increase of staining density in the lateral direction and then an abrupt decrease of staining exactly at the lateral limit of the layer. As a result, the characteristic, rounded, dorsolateral end of layer V forms a conspicuous landmark, equally distinct in all planes of sectioning (Vdl, Figs. 14a–17a, 29a–31a, 42a–45a).

Layer VI, as defined on the basis of Golgi investigations, is difficult to separate from layer V with most other methods but may be differentiated with the stain for α-glycerophosphate-dehydrogenase (Mellgren and Blackstad, 1967, Fig. 9). The sulphide silver staining of neuropil and cell bodies is gradually reduced near white matter in a zone corresponding to layer VI (VI, Figs. 17a, 19a, 21a, 22a, see also 30a–31a). Thus defined , layer VI can often be followed right out to the lateral border of area 28M, although this is not evident in the present figures.

A medial border of area 28M is not discernible in layers V–VI. As mentioned, the sulphide silver pattern suggests that there is one continuous lamina interna belonging to the entorhinal area, the parasubiculum and the presubiculum. Within this layer the staining intensity decreases gradually from fairly strong dorsolaterally (Figs. 14a–17a) to weak medially (Figs. 16a–20a)–and to virtually none ventromedially (Figs. 25a–26a; see also Figs. 32a–36a).

b) Lateral Part, Area 28 L (Figs. 18a–24a, 30a–35, 41a–44a)

Layer I has a pale superficial and a dark deep sublayer which are continuous with the corresponding sublayers of area 28M (Is and Id, Fig. 21a). The superficial sublayer stains somewhat more in area 28L (IsL, Fig. 21a, 32a) than in area 28M (IsM, Figs. 21a, 32a). Furthermore at rostroventral levels it has an *unstained zone subpially* which is rather thin but conspicuous for its sharp delimitation (between z and z, Figs. 22a–23a, 31a–34a). This subpial white zone gradually becomes thicker as it is followed rostralwarts through the transitional zones and into the pyriform cortex of which it is a general characteristic (Figs. 24a–and 35a–).

The dark staining which defines the deep sublayer of I continues into layer II.

Layer II. This, together with the deepest part of I, stains darkly. The resulting broad dark zone differs strikingly from the corresponding layers in area 28M (compare them in Figs. 20a–21a or 30a–32a). It is somewhat patchy dorsomedially (near to area 28M, e.g. in Figs. 30a–31a) due to deep indentations of its superficial and deep borders. The deep border of the dark stain is not quite sharp; it correlates roughly with the cytoarchitectonic border between layers II and III, which itself is somewhat poorly defined.

The somata of the outer sublayer (cp. cytoarchitectonic description) are unstained, while those of the inner sublayer frequently bear a dark granular staining as in layer II of area 28M. Clusters of stained somata often coincide with the dark patches of neuro-pil-stain (not illustrated here).

Layer III has the same intermediate staining density as in the medial part.

Layer IV lacks the sharp definition that is seen in area 28M since the superficial and deep pale lines are lost in the transition into area 28L. Nonetheless the layer is usually identifiable by its slightly enhanced staining relative to layers III and V (Figs. 19a–23a, 31a–32a) and, in lightly stained sections, by lack of staining in its somata (Figs. 42a–44a).

Layers V and VI are inseparable, staining roughly as layer V of area 28M.

In sulphide silver stained sections, the borders of area 28L are not sharp compared to, e.g., the dorsal, dorsolateral or medial limits of area 28M. However, the transition dorsolaterally (cp. Figs. 3, 6, 10) into the perirhinal area (Krieg, 1946) is quite abrupt (Figs. 19a–22a, 32a–35); taking place either just deep to or a little caudoventral to the rhinal fissure. Here the dark stain, covering entorhinal layers I (inner part) and II, expands lateralwards to make one broad dark zone which spans perirhinal layers II–V. Layer III of area 28L is replaced, as it were, by this dark neocortical zone. The slightly darker staining in layer IV of area 28L is lost at the interareal border (arrows, Figs. 21a, 22a). Lamina interna of area 28L continues into an intermediately stained zone just above the neocortical white matter (broken arrow, Figs. 20a–23a).

c) Transition between Medial and Lateral Parts (Figs. 18a–22a, 29a–32a, 41a–44a)

The most marked changes in the sulphide silver pattern between areas 28M and 28L are the following:

1) The staining of layers I/II increases gradually in the mediolateral direction (Figs. 20a–21a, 31a) and most of this change takes place within a fairly well defineable transition area (Figs. 18a–23a, 30a–32a, 41a–44a). For the purpose of plotting, the medial starting point of this gradient was taken where the staining density just begins to increase above that of layers I/II further dorsomedially (2', Figs. 18a–23a, 30a–32a, 41a–44a). Some isolated patches of dark stain (e.g., at arrow labelled with asterisk, Fig. 19a) are often seen medial to this border, as one might initially identify it, and when it seemed reasonable, the border was modified to include these patches in the transition zone (Fig. 20a). Similarly an end-point to the gradient was taken where the staining density appeared to have reached a plateau (2, Figs. 18a–22a, 30a–32a, 41a–44a). The resulting transition zone extends somewhat further medially than the cytoarchitectonic transition in layer II (Figs. 4, 9).

2) The blurring of the pale lines delimiting layer IV is more abrupt than the change of the staining in layer II. Thus it suffices to indicate only one transitional point per section in layer IV (filled circles, Figs. 18a–23a, 31a–33a), and these points (filled circles, Figs. 1, 4, 7) define a fairly smooth curve, suggesting good reliability. It is seen that the border in layer IV (filled circles) runs approximately along the lateral side of the transition zone stippled in layer II (Figs. 1, 4, 7).

3) The sharply delimited lateral end of layer V (typical of area 28M) is seen to disappear between the horizontal levels of Figs. 17a and 18a, which is the level where area 28L appears, as judged by the other criteria.

4) As in the cytoarchitectonic pattern, the lateral limit of lamina externa changes gradually from the appearance characteristic of area 28M to that characteristic of area 28L.

d) Ventrolateral Part, Area 28L' (Figs. 24a–25a, 34a–37a, 41a–44a)

There are gradual changes in sulphide silver pattern, more or less paralleling the cytoarchitectonic ones, between area 28L and adjoining well defined areas. We will now describe the transition to a) the pyriform cortex and b) area TR in parallel with the description of the cytoarchitectonic changes given above.

α) *Laterally in Area 28L*, starting at the border towards area 28M and ending in the pyriform cortex, we encounter the following changes:

1) Gradual, moderate, darkening of the staining in the outer half of layer I (Figs. 20a–24a, 31a–39a, 40a, 41a). 2) Appearance (Fig. 21a, 31a) and broadening (Figs. 22a–24a, 32a–39a) of the narrow unstained subpial zone. 3) Blurring of the deep border of layer II and darkening of III (Figs. 20a–25a, 31a–38a). This results in the characteristic Timm pattern of the pyriform cortex, with maximal staining in the neuropil just superficial and deep to layer II and moderately decreased staining towards the depth of layer III (Fig. 39a). 4) Layer IV (IV, Fig. 33a) gradually becomes indiscernible (Figs. 34a–37a–39a). 5) The intermediate staining of lamina interna (V–VI, Figs. 34a–35a) may be traced (through Figs. 36a–37a) into the posterior pole of the claustrum (cl, Figs. 27, 38a–39a, see also Fig. 41a,b).

β) *Medially in area 28L* the following changes occur rostrally, towards the transition area, TR:

1) Darkening of the stain in the outer half of layer I–as above. 2) The subpial white line spreads medialwards in area 28L and further into area 28M' (Figs. 23a–24a). 3) Dark staining spreads rapidly from layer II into III. Thus it is possible to indicate a border, rostral to which layer III is darkly stained (black diamonds, Figs. 4–10, 23a, 33a, 42a–47a). 4) Just a little further rostroventrally layer IV suddenly is obscured by patches of dark staining which also involve layer V and thus go nearly all the way to the white matter (between y and y, Figs. 24a–26a, 34a–36a, 41a–42a). This area of patchy staining encircled by curve y (Figs. 3, 4, 6, 7, 9, 10) partly overlaps with what was cytoarchitectonically defined as area 28L' and partly with the area TR.

Thus it is seen that the gradual changes that develop *laterally* in area 28L, (over the levels of Figs. 34–39, stippled zone Figs. 6 and 7) and *medially* in area 28L (over the levels of Figs. 33–35) only roughly parallel the cytoarchitectonic changes described above. The areas or zones where the transitions occur, as defined by different criteria, are not sharply circumscribed and coincide only in part with each other.

e) Ventromedial Part, Area 28M' (Figs. 24a–27a, 33a–36a, 45a–47a)

At the ventral end of area 28M, the following changes occur in the sulphide silver staining pattern over a range of some 500–800 µm:

1) and 2) as above.

3) The staining of the medial end of layer III becomes lighter so that the difference in density between the medial ends of layers II and III, characteristic of most of area 28M, is no longer marked (Figs. 22a–23a).

4) The dark staining which "spreads" from layer II into layer III further dorsolaterally (see description of area 28L' above) also expands to fill the combined layers I (inner zone)/II/III throughout the lateromedial extent of area 28M' (Figs. 24a–27a, 33a–36a and 45a–47a).

5) The narrow lamina dissecans (IIIa and/or IV) is at first conspicuous in area 28M' (Figs. 24a, 34a), its superficial limit being further enhanced by the very dark staining in the lamina externa. Further rostroventrally in area 28M', laminae dissecans and interna become inseparable, as the staining of lamina interna is reduced (Figs. 25a–27a, 34a–36a). The lateral end of the pale layer IV forms a fairly well defined border (filled circles Figs. 1, 4–7) between areas 28M' and either 28L' (Figs. 24a, 34a–35a) or TR (Figs. 25a–27a).

As previously mentioned, lamina interna becomes gradually paler at more ventral levels of area 28M and especially within area 28M'. There it becomes continuous with the equally pale deep zone of the cell layer in the ventral part of the subiculum (Figs. 24a–27a, 34a–36a).

In area 28M' (Fig. 24a) lamina externa as a whole, and especially a superficial part of layer I, is darker than in the ventral tip of the parasubiculum (Fig. 23a, see also 34a–35a). This facilitates the separation of the two areas.

f) Transition Area, TR

A small caudal part of the transition area, TR, shows the patchy sulphide silver staining described above (circumscribed by interareal border y, Figs. 3, 4, 6, 7, 9, 10, 25–26, 36). The remaining rostral part of area TR is more evenly stained (Figs. 27–28, 37–38).

IV. Discussion

Before considering the significance of the present observations, it is necessary to emphasize a few considerations which have been fundamental to the present study.

Laminae and areas that represent fields of origin or termination of specific fibre systems are particularly important within any subdivision of the cerebral cortex, connectivity being basic to nervous function. However, experimental methods for mapping connections are time consuming and therefore best directed at specific problems. Necessary reference maps for locating lesions, labelled fields, terminal degeneration, etc., are therefore usually based on cyto- and fibre-architectonics or—recently—chemo-architectonics. Histochemical methods that give clear and reproducible staining patterns may be very useful for this purpose even before their histochemical specificities have been determined or the exact morphological structures which they stain have been identified, as is presently the case with Timm's method.

On the other hand, to determine the functional significance of this staining pattern it is obviously necessary to know its chemical and cellular basis. While, the sulphide silver stainable metals and metal containing substances have not yet been identified, evidence cited in the introduction shows that the stain in the neuropil must depend on certain fibre systems and have, wholly or in part, a presynaptic localization. Laminar and areal differences in this staining are therefore obviously significant criteria for subdividing the cerebral cortex. However it remains to be determined *which* fibre systems are the Timm positive ones in the parahippocampal cortex. The following discussion therefore falls into two parts.

First, it is discussed to what extent the Nissl and Timm patterns confirm and improve current views on parahippocampal parcellation when the two methods are used simply for differentiating laminae and areas, and how far current knowledge on connections supports the present subdivisions. Afferent fibre systems are in this context related to areas rather than to the sulphide silver pattern per se.

Second, it is briefly considered whether the normal staining pattern offers any clues to the identity of the particular fibre systems—or of the particular chemical components—that are stained with the sulphide silver method within the parahippocampal areas.

A. Regional and Laminar Subdivision

1. Parahippocampal Areas except the Entorhinal Area

a) Presubiculum

The present observations agree fully with the current definition and subdivision of this area, developed gradually through studies of internal structure (Cajal, 1911, 1955, Lorente de Nó, 1933), connections (Blackstad, 1956) and chemo-architectonics (Storm-Mathisen and Blackstad, 1964; Mellgren and Blackstad, 1967).

External borders. These were previously well established and are independently confirmed by the sulphide silver pattern. As in previous morphological and histochemical studies, only the lamina externa shows clear interareal borders between the presubiculum and the parasubiculum.

Subareas. The presubiculum shows clear dorsoventral differences in thickness, width and position relative to the hippocampal fissure, and some simplification of the laminar structure at far ventral levels. Since these changes are very gradual, they will not be used here for defining dorsal and ventral subdivisions, like areas 27a,b and 48 of Brodman (1909) or Prsub 1 and 2 of M. Rose (1929, mouse; 1931, rabbit; see further discussion by Stephan, 1975, pp. 727–728). Other recent reports have taken the same point of view (Storm-Mathisen and Blackstad, 1964; Geneser Jensen and Blackstad, 1971).

A subdivision of the presubiculum into zona medialis and zona lateralis is proposed by Stephan (1975, pp. 722, 728) in order to take into account the pronounced island formation medially in layer II, particularly of humans, and the interdigitation of the subicular and presubicular cell layers at the border between these two areas. In the rat, the island formation in layer II is slight, however, and the zone of overlap between the presubiculum and the subiculum is so narrow (see Blackstad, 1956, Fig. 7; Hjorth-Simonsen, 1973, p. 146 and Figs. 2a, 5; Haug, 1974; and present illustrations), that it may not require a separate name.

Laminae. It is probable that the present cytoarchitectonic laminae correspond to those defined by Cajal (1911, 1955) although there are some discrepancies.

Thus, Cajal's layers b and F (Cajal, 1911, Fig. 449, mouse) which he calls 2 and 3 in the text, correspond approximately to the present layers II and III in relative thickness. However, in some of Cajal's other drawings the relative widths of his layers are different from what is seen in the present material. His layer 2 (Cajal, 1955, Fig. 16, guinea pig) or B (Cajal, 1911, Fig. 448, man; 1955, Fig.

14, man) is thicker than the present layer II and thinner than the present layer III. It should probably be equated with the present layer II. Cajal's layer 3 (Cajal, 1955, Fig. 16, guinea pig) or C (Cajal, 1955, Fig. 14, man) is thick but poor in cells and has a massive fibre plexus. Although the present layer III can hardly be called cell poor, it is at least less compact than layer II and is the likely equivalent of Cajal's layer 3. The present layer IIIa is thin and somewhat inconstant and was apparently not found worthy of distinction by Cajal (1911, 1955). Cajal's layer 4 (Cajal, 1955, Fig. 16, guinea pig) or D (Cajal, 1955, Fig. 14, man) may correspond to the present IV, although the latter again is somewhat thinner than Cajal's layer. Cajal's layer E (Cajal, 1911, Fig. 448, man; 1955, Fig. 14, man) and 5 (Cajal, 1955, Fig. 16, guinea pig) correspond to the present combined V/VI. The difference in width between the present and some of Cajal's layers in the presubiculum does not merely represent species differences. A recent study of the Timm and Nissl patterns in the guinea pig (Geneser Jensen et. al., 1974) shows the same relative thickness of these layers as in the present rat material.

The Timm and Nissl patterns agree as concern interlaminar borders. In addition, the sulphide silver pattern enhances layers II and IV. It should be remembered here that sulphide silver staining of somata contributes significantly to the over all density of layers II, III and IV of the presubiculum. This complicates the study of the moderately stained neuropil. Nevertheless, at least some difference in staining density of the neuropil is discerned at all major interlaminar borders.

Laminae I, II and III, as recognized here, are revealed also with reduced silver methods (Blackstad, 1956, see also Fig. 2 in Mellgren and Geneser Jensen, 1972; White, 1959) or by staining for acetylcholinesterase (Storm-Mathisen and Blackstad, 1964) or some oxydoreductases, particularly α-glycerophosphate dehydrogenase (Mellgren and Blackstad, 1967; Mellgren, 1973a). With the last mentioned method lamina dissecans is in addition definable.

b) Area 29e

Definition and external borders. An area 29e was described by Brodman (1909) and Zunino (1909) in the rabbit. M. Rose (1931) and Rose and Woolsey (1948) who studied the same species, included this area into the presubiculum.

A smaller and more ventrally situated area 29e was then defined in the rat by Vaz Ferreira (1951) and Blackstad (1956). This area, in the rat, is characterized by lacking the prominent fibre plexuses present in the adjoining pre- and parasubiculum. Other distinctive features of the rat's area 29e include commissural innervation, which is absent at corresponding dorsoventral levels of the presubiculum (Blackstad, 1956), and a lack of cingulum innervation, which is present in the pre- and parasubiculum (White, 1959). On the other hand, the difference in connections is not total, since presubiculum's *layer II* resembles area 29e in showing sparse cingulum degeneration (White, 1959) and weak cholinesterase staining (Storm-Mathisen and Blackstad, 1964), and since commissural degeneration in the parasubiculum may merge with that in area 29e (Blackstad, p. 479).

In a study of normal, fibre impregnated, material Smith and White (1964) included a similar fibre poor area in the cat's parasubiculum, a view which was recently accepted by Stephan (1975, pp. 721—722). In the guinea pig, on the other hand, no area comparable to the rat's area 29e could be found with acetylcholinesterase or sulphide silver staining.

The strong sulphide silver stain defines the rat's area 29e more sharply in normal material than do any other methods so far employed for this purpose. This should facilitate the comparative anatomical study of this area.

The term area 29e implies similarity to other retrosplenial areas. However, as pointed out by Blackstad (1956) and White (1959) there are no data on structure or connections to support this relationship, nor does the topography particularly suggest it. Topographically, the relation with pre- and parasubiculum is closer. Particularly the continuity of the sulphide silver stained area 29e with layer II of the presubiculum could imply a close relation with the latter. However, the data just cited show that area 29e differs from both the presubiculum (i.e., in receiving commissural afferents) and the parasubiculum in structure and afferent connections. It therefore seems most practical, at present, to retain a separate name for this area.

Laminae. Cytoarchitectonic lamination in the present material is consistent with earlier descriptions and is supported by the sulphide silver staining, although the latter encroaches slightly on the molecular layer. Whether or not to include a sector of laminae dissecans and interna in area 29e remains arbitrary.

c) Parasubiculum

The present observations support the current definition and subdivision of this area in the rat (Lorente de Nó, 1933; Krieg, 1946; Blackstad, 1956; Storm-Mathisen and Blackstad, 1964; Mellgren and Geneser Jensen, 1972; Mellgren, 1973a,b).

External borders. Cytoarchitectonically, these are fairly well defined. They are further enhanced with the sulphide silver method—except for the medial border at the ventral extreme of the area.

Subareas. The distinction *between subareas a and b* is uncertain when considering the Nissl material alone. It was first made by Lorente de Nó (1933) in the macaque but not in the rat, and was ignored by Krieg (1946). Staining for acetylcholinesterase, oxydative enzymes or monoamine oxydase in the adult or developing rat brain gives at best only vague indications of the subdivision into subfields a and b (Storm-Mathisen and Blackstad, 1964; Mellgren and Blackstad, 1967; Mellgren and Geneser Jensen, 1972; Mellgren, 1973a,b). However, the subareas are clearly defined in the rat by the presence in b and absence in a of a normal fibre plexus and of commissural degeneration (Blackstad, 1956; see also Fig. 2 in Mellgren and Geneser Jensen, 1972, for illustration of the normal fibre plexus). Again, the sulphide silver method shows a clear difference in staining density and a sharp border between the two subfields. This border follows the fibrearchitectonic border quite closely, perhaps without coinciding with it completely. A possible discrepancy is that parasubiculum a may expand more at ventral levels when defined fibrearchitectonically than when defined with the sulphide silver method. Whether this is so can only be settled in adjacent sections stained with the two methods.

The tendency for the sulphide silver stain to increase somewhat towards more ventral levels of parasubiculum b is consistent with the tendency for the normal fibre plexus of this part to thin out gradually at far ventral levels (Blackstad, 1956).

The small patch of darker sulphide silver stain medially in area 49b at ventral levels is not related to any other known chemical or structural specialisation. It is remarkable that such a small patch is always present.

It is interesting that the guinea pig parasubiculum can not be divided into subareas a and b with any morphological or histochemical methods so far applied to it (Geneser Jensen et al., 1974, and references therein).

Laminae. The sulphide silver pattern confirms the cytoarchitectonic lamination which again is consistent with previous accounts (cited above). At ventral levels individual laminae are gradually reduced in thickness and the cytoarchitectonic laminar scheme is simplified by the disappearance of layer IV. These changes paralell the architectonic gradients seen in the presubiculum and (see below) the entorhinal area.

2. Area Entorhinalis and Transition to the Pyriform Cortex and the Amygdala

Ambiguities have arisen as authors have based their subdivision of the entorhinal area on a single principle, particularly cytoarchitectonics; have used a single plane of sectioning or too widely spaced sections; or have ignored ventral parts of the area (see Appendix). The gradual medio-lateral and caudorostral changes that exist within this region have contributed to the difficulties in subdividing it. The present study attempted to circumvent at least some of the above limitations. However, as the borders traced in Figs. 3–10 do not fall into groups of completely coinciding curves, there is no obviously correct way of subdividing the area on this basis alone. Accordingly, it would not be surprising if new methods were to reveal yet other borders, incongruent with those shown here. It also seems quite probable that many interareal borders vary between individual rats, a possibility which may relate particularly to zones of gradual changes and non-coinciding borders.—However, a distinct and quite invariable border between areas 28M and 28L was found in the guinea pig (Shipley et al., 1974).

Fig. 49 sketches a subdivision which seems reasonable at present.

a) Areas 28M and 28L

These areas were already firmly defined by cytoarchitectonics, fibrearchitectonics, chemoarchitectonics and connections (Cajal, 1911, 1955; Lorente de Nó, 1933, 1934; Blackstad, 1956, Hjorth-Simonsen, 1971, and others—see Appendix). However, in the present study the extent of these areas is systematically mapped, and new criteria are given for defining their ventral limits. Moreover, previous suggestions of a transitional zone between areas 28M and 28L (Lorente de Nó, 1934; Hjorth-Simonsen, 1972; Van Hoesen and Pandya, 1975a,b; Van Hoesen et al., 1972, 1975) are supported by the present observations. The cytoarchitectonic transition zone, as recorded here, is narrow but not insignificant, and the changes in the sulphide silver pattern, occurring more medially, broaden the zone where characteristics of areas 28M and 28L intermingle. This composite transition zone probably includes all those borders and transitions that have been described between areas 28M and 28L with other methods (cited above) and may correspond at least roughly to Lorente de Nó's (1934) part B. However, it is not yet supported by any systematic study of afferent and efferent connections in the rat, although Hjorth-Simonsen (1972) reported a single observation strongly suggestive of an intermediate perforant path.

Laminae. The interlaminar borders already established in areas 28M and 28L of the rat (and of the guinea pig, Geneser Jensen et al., 1974) are also confirmed by the sulphide silver method. However, the pattern of lightly and darkly stained zones suggests termination of yet other fibre systems (extrinsic or intrinsic), and the existence of yet other differences in afferent connections between areas 28M and 28L, than presently recognized.

b) Areas 28M', 28L' and TR

These do not correspond precisely to any previously well defined subareas, although the heterogeneity of this region has often been described, and occasionally put into maps that are somewhat similar to those presented here (see Appendix).

Area TR is here defined cytoarchitectonically rather than by its sulphide silver staining. The latter, being patchy caudally and evenly distributed rostrally, is neither uniform within nor unique to the area TR. But the very characteristic single cell layer of this area amply justifies separating it from the surrounding areas. It was described, under a different name (ed2 and ed3) by the Popoffs (1929); it probably corresponds to the cortical amygdaloid nucleus of Krieg (1946) and was described but left unnamed by Price (1973). It really bears no cytoarchitectonic resemblance either to other parts of the entorhinal area or to the pyriform cortex. The fact that it becomes continuous rostrally with what is often called the corticoamygdaloid transition area (area semiannularis , Pam Cs, of Stephan, 1975) suggests that it may be included as a posterior part of the latter. However, the corticoamygdaloid transition area is not a very well defined term, so this possibility is better considered in connection with the pyriform cortex and the amygdala.

Area 28M' differs markedly from areas 28M and 28L and is fairly homogenous, both cytoarchitectonically and with the sulphide silver method. Despite its lack of a really sharp posterior border it is therefore defined here with assurance. Previous accounts have not clearly recognized it as a separate subarea.

Area 28L'. The observations presented above support Price and Powell's (1971) and Price's (1973) descriptions of gradual changes in structure between the sphenooccipital ganglion of Cajal (present area 28M), and the pyriform cortex, and at the same time confirm that most of the original entorhinal-pyriform transition area of Price and Powell (1971) should be identified as area 28L in the sense of Blackstad (1956).

The present "area" 28L' lies rostral to area 28L, and is defined as the transition zone between the more characteristic areas 28L, TR and the pyriform cortex. It is sufficiently heterogeneous to invite further subdivisions into a medial part (lying between areas 28L and TR) and a lateral part (forming a smooth transition between area 28L and the pyriform cortex) and if this is done, we shall get a map *in principle* more similar to those of Rose (1929, 1931) and Popoff and Popoff (1929). However, the resulting subareas would be small, vaguely defined and therefore of little practical value if not supported by additional data on structure and connections. In fact, the definitive recognition of area 28L' must await such data.

It is appropriate to ask whether the suggested subdivision of the ventral part of area entorhinalis is consistent with its known *connections.*

It is now well established that secondary olfactory fibres reach part of the entorhinal. area but not area 28M as defined in the present paper (White, 1965; Scalia, 1966, Heimer, 1968; Kerr and Dennis, 1972; Price, 1973). Some uncertainty remains as to whether the whole or only a ventral part of area 28L receives secondary olfactory afferents. Therefore, the white subpial line in sulphide silver stained material is interesting. As discussed below it may represent the terminal field of secondary olfactory afferents and if so the latter often fails to reach the caudal third of area 28L. In any event the distribution of secondary olfactory afferents, by involving all three subareas 28M', 28L' and TR, neither confirms nor contradicts the present definition of these areas. Moreover, by extending into area 28L it fails to provide information on the separation of area 28L' from area 28L. Fibres from the pyriform cortex apparently follow the areal distribution of secondary olfactory afferents within the entorhinal area although they end at complementary dendritic levels (Price, 1973).

On the other hand, fibres from the amygdala may have a differential distribution respecting the subdivisions proposed here. In an initial study of these connections, Krettek and Price (1974) reported projections from the rostral part of the *cortical amygdaloid nucleus* to layers I and II of the "anteromedial tip of the enthorhinal cortex" (present area 28M'?) and from the *basolateral and lateral nucleus* to the "lateral and ventral portions of the entorhinal cortex" (present area 28L'?).

Efferent connections of the ventral parts of the entorhinal area have not been specifically studied. Hjorth-Simonsen (1972) could not decide whether the transitional zone of Price and Powell gives rise to perforant path fibres and the situation is the same with regard to area 28L'. In fact it may be of interest to determine whether the field of origin of the perforant path (lateral perforant path in this case) is sharply delimited against the pyriform cortex or if there is a transition in the form of a gradually decreased proportion of somata that give off perforant path fibres. This possibility was not discussed by Segal and Landis (1974).

It should also be determined whether area TR gives off perforant path fibres. Fig. 4c (Lesion No Si 77) of Hjorth-Simonsen (1972) suggests that this is not so.

c) Comment on Area Perirhinalis

Van Hoesen et al. (Van Hoesen and Pandya, 1975a,b; Van Hoesen et al., 1972, 1975) recently advanced the interesting suggestion that what they call the perirhinal and prorhinal areas on the lateral, respectively medial, banks of the rhinal fissure are relay zones between the neocortex and the enterhinal area.

As discussed in the Appendix, these authors' prorhinal area 1 seems at present not well defined. Their prorhinal area 2, however, is rather extensive and should be looked for in lower species. The present material does not reveal it. Whether fibre architectonics or renewed studies on afferent and efferent connections will do so, remains to be seen.

The perirhinal area would lie just lateral to areas 28M and 28L as defined here. Its *lateral* border appears ill defined cytoarchitectonically (see also Krieg, 1946, Fig. 12), and in the sulphide silver pattern when compared with the many sharp interareal borders seen *within* the hippocampal region. For a detailed discussion of the literature on area perirhinalis, see Stephan (1975, pp. 648–649, 666, 678–679, 687, 715)

d) Concluding Remarks on Entorhinal Subareas

Lorente de Nó (1934) postulated that there is a fundamental plan of the entorhinal area, which is reflected in its mediolateral subdivision and remains constant at all dorsoventral levels within a species and throughout the phylogenesis. Although admitting the presence of dorsoventral changes within the area, he strongly disagreed with authors of the cytoarchitectonic school, who stressed the dorsoventral changes and put forward complicated schemes of subdivision that differed considerably from one species to another (see Appendix). Lorente de Nó's just criticism of inconsistencies in the cytoarchitectonic subdivisions together with his emphasis on the mediolateral subdivision may have caused later authors to neglect the dorsoventral differences. On the other hand, his emphatic pointing out that the structure of this area changes *gradually* rather than abruptly in the mediolateral as well as the dorsoventral sense was not followed up.

Now, the distribution of acetylcholinesterase (Geneser-Jensen et al., 1971; see also Fig. 10 of Hjorth-Simonsen, 1971), certain afferent fibres (Krettek and Price, 1974) and the sulphide silver staining (Geneser-Jensen et al.,1974 and present observations) all indicate that the dorsoventral difference in cytoarchitecture are correlated to differences in afferent—or internal—connections. In addition, a *gradual* change is found between areas 28M and 28L and, finally, the present study reveals mediolateral gradients even *within* these areas. For a definitive subdivision of the entorhinal area to be achieved, these data must be correlated and supplemented. To describe reliably the relations between the many architectonic gradients that characterize this region may, however, require not only a graphic but a quantitative and statistic approach.

Stephan's (1975) treatise on the allocortex became avaliable as the present manuscript was about to be submitted for publication. His careful and complete discussion of the literature to about 1974 covers all morphological aspects of the parahippocampal cortex. Where it overlaps the present discussion there appears to be agreement on all major points—disregarding differences in terminology.

Although Stephan (1975, pp. 647—648, 676) accepted the principal subdivision of the entorhinal area into medial and lateral parts in lower mammals, he stressed the want of reliable criteria for subdividing the more differentiated entorhinal area in primates and for establishing homologies between primates and the better investigated lower animals (Stephan, p. 676—679). The need for detailed comparative anatomical investigations of this region is reinforced by the definition of additional subareas in the rat; the more so since cytoarchitectonic studies (see Appendix) suggests as much differentiation within the entorhinal cortex of other lower mammals.

B. Are the Sulphide Silver Stainable Structures and Substances Identical to Previously Identified Tissue Components?

Subdivision of the parahippocampal region was the focal point of the previous section. Thus we have already touched on the relations of the sulphide silver staining to specific fibre systems and to specific enzymes.

Now we may consider this question more directly. Pending experimental evidence, a close similarity between the distribution of sulphide silver staining and the distribution of particular fibre systems—or of some other histochemical reactivity—might serve to identify the stainable tissue elements.

This approach has several pitfalls. Thus, the distributions to be compared may represent more than one type of elements, or correlations may reflect indirect associations, such as two fibre systems having identical terminal distributions when seen in the light microscope. Nevertheless, in the hippocampal formation striking similarities between the distribution of sulphide silver staining and the terminal distribution of particular fibre systems led to a working hypothesis on the identity of the Timm positive fibre systems (Haug, 1974) which has been partly supported by later experimental work (Haug, 1976 and in preparation).

The parahippocampal areas have so far been less carefully charted with respect to the detailed laminar and areal distribution of afferent fibre systems than the hippocampal formation. Therefore, definite suggestions on the identity of the Timm positive fibre systems are hardly warranted here. Most of the afferents to these areas end in moderately to strongly sulphide silver stained laminae and may therefore well have sulphide silver stainable boutons.

The terminal fields that have been mapped most carefully are probably those of commissural fibres (Blackstad, 1956, Fig. 6) and they do not appear to be closely associated with the sulphide silver staining. Nevertheless, we may not strictly exclude that some—or even that all—of the commissural fibres are Timm positive.

Other afferents that could be Timm-positive are cingulum afferents (White, 1959; Domesick, 1970, 1973; Shipley and Sørensen, 1975), at least the fibres ending in layer III; fibres from the presubiculum to area 28M shown in the guinea pig (Shipley, 1975); fibres from CA3 to area 28M—and area 28M'? (Hjorth-Simonsen, 1971); fibres from the pyriform cortex (Price, 1973) and the amygdala (Krettek and Price, 1974), and possible counterparts to the neocortical afferents recently discovered in the monkey (Van Hoesen and Pandya, 1975, Van Hoesen et al., 1972, 1975). Lorente de Nó (1933) described three types of afferents to the entorhinal area in Golgi material. One or more of these types could be Timm-positive. On the other hand, largely unknown fibre systems may well cause the neuropil stain and we should remember that probably only a minority of the terminals in any layer needs to be Timm positive in order to cause heavy staining in thick sections.

On the present evidence, secondary olfactory afferents are likely to be Timm negative since these fibres end in layer Ia (Price, 1973) which again corresponds closely to the subpial, completely Timm negative, zone throughout the pyriform cortex and the entorhinal area.

Histochemical maps sufficiently detailed to be of interest here are those for acetylcholinesterase (Storm-Mathisen and Blackstad, 1964), various oxydoreductases (Mellgren and Blackstad, 1967), monoamine oxydase (Mellgren and Geneser Jensen, 1972) and catecholamines (Blackstad, et al., 1967). None of these distribution patterns seem very closely related to the sulphide silver pattern. The distribution of α-glycerophosphate dehydrogenase (Mellgren and Blackstad, 1967, Fig. 9) is to some extent an exception; it does resemble the sulphide silver pattern in area 28M—and on some points within the hippocampal formation itself (see illustrations in Haug, 1974). On the other hand it differs from the sulphide silver pattern in the staining of the parasubiculum and on other points within the hippocampal formation. Therefore the present material does not strongly suggest that α-glycerophosphate dehydrogenase contributes to the sulphide silver pattern.

V. Summary

This study has the twofold purpose of revising the current subdivision of the rat's para-hippocampal cortex (comprising presubiculum, area retrosplenialis e, parasubiculum and area entorhinalis), and of mapping the terminal distribution of sulphide silver stainable fibre systems within this region.

A combined chemo- and cytoarchitectonic atlas, based on closely spaced horizontally, frontally and parasagittally cut serial sections, is presented. It is supplemented with graphically reconstructed surface maps of areas, interareal borders and transitional zones. The comparison of the chemoarchitectonic (i. e., sulphide silver-) and cytoarchitectonic patterns leads to the following conclusions:

Previously defined interareal and interlaminar limits of the presubiculum, area retrosplenialis e and the parasubiculum, and previously defined interlaminar limits within areas 28M and 28L of the entorhinal cortex, are confirmed in both staining patterns, but are revealed most clearly and completely with the sulphide silver method. Thus, the latter, being independent of any of the other histological methods used for subdividing this cortical region, provides strong evidence for the importance of the previously defined laminae and areas.

Previous suggestions that there is a transition rather than a sharp border between areas 28M and 28L are supported by both staining patterns.

Rostroventrally in the entorhinal cortex, many features change gradually as the pyriform cortex is approached, and borders and transition zones seen with the two methods and in different layers rarely coincide. This emphasizes earlier reports of a gradual transition between the entorhinal and pyriform cortices, and presents obstacles to any attempt at delimiting and subdividing this part of the entorhinal cortex. Thus, sharp interareal borders cannot be drawn here. Nevertheless it is possible to define three new ventral subareas, i. e., the ventromedially situated area 28M', the ventrolaterally situated area 28L', and area TR which is located between the latter two areas, the amygdala and the pyriform cortex. Area 28M' has a unique cytoarchitecture as well as sulphide silver pattern; area TR has a unique cytoarchitecture but is heterogeneous in the sulphide silver pattern; area 28L' has a cytoarchitecture and a sulphide silver pattern intermediate between area 28L, area TR and the pyriform cortex.

A selective review of earlier subdivisions of the entorhinal area is given in an appendix.

The sulphide silver pattern is of interest not only in the context of entorhinal subdivisions. Since previous and ongoing studies suggest that the sulphide silver staining in the forebrain neuropil resides in synaptic boutons or associated postsynaptic structures, the present detailed description of its distribution in the parahippocampal cortex may be taken to reflect the form and density of the terminal fields of "sulphide silver positive fibre systems", whose origin and course remain to be determined. Experimental studies with this purpose may be based on the present description of the normal staining pattern.

Appendix

Review of Previous Subdivisions of the Entorhinal Area

Cajal (1911) recognized the following five regions in the hippocampal gyrus of man and its equivalent, the pyriform lobe, of small mammals:

  I: Region olfactive principale (1911, p. 686)

  II: Subiculum (1911, p. 698)

 III: Presubiculum (1911, p. 703)

 IV: Region externe ou fissuraire de l'hippocampe (1911, p. 705)

  V: Ecorce temporal postérieure ou supérieure; centre ou noyau temporal supérieure, noyau angulaire (1911, p. 706).

The original papers (Cajal, 1901-1902, 1955) recognized the same subdivisions under slightly different headings (page references to Cajal, 1955):

Hippocampal Convolution (Man)

A. The subiculum (p. 27). The cornual or subicular portion (p. 28).–Equivalent to II above.

B. The central or salient portion of the gyrus (p. 27). The central cornual portion (p. 29).
   a) Presubicular region (p. 29).–Equivalent to III above.
   b) External, or olfactory portion of the central cornual region (p. 30). Presumably also: Olfactory sphenoidal cortex (legend to Figs. 22, 23), sphenoidal cortex (central external region) (p. 43). sphenoidal cortex (legend to Fig. 36), spenoidal olfactory cortex (legend to Fig. 108).–Equivalent to I above.

C. The lateral external portion bordering the limbic or rhinal fissure (p. 27). External, or fissural portion of the hippocampal convolution (p. 31).–Equivalent to IV above.

The pyriform (or sphenoidal) lobe of mammals (pp. 31–32) was said to have essentially the same structure as the hippocampal convolution.
Special ganglion of the spheno-occipital cortex; Angular (spheno-occipital) ganglion (p. 164). Equivalent to V above. The actual descriptions of the spheno-occipital ganglion in both texts (Cajal, 1911 and 1955) seem to refer only to small mammals and not to man.

In distinguishing between I and V above, Cajal included the ventral parts of area 28 into his *Region olfactive principale.* The mediolateral relation that he gave within the hippocampal convolution and the pyriform lobe show this. He stated that the presubiculum is followed laterally by the *Region olfactive principale* (Cajal, 1911, pp. 684–686; 1955, pp. 27, 31); a location taken up by parts of area 28 of later authors (Brodman, 1909; Rose, 1926, 1927, 1929, 1931; Popoff and Popoff, 1929; Meyer and Allison, 1949; Allison, 1954; present maps). Even the present area 28L probably lies within Cajal's *Region olfactive principale* since area 28M was originally defined as having the structure of the *spheno-occipital ganglion* (Blackstad, 1956; Scalia, 1966; Kerr and Dennis, 1972). It is true that in Cajal's view (1955, first paragraph p. 177) the spheno-occipital ganglion and the sphenoidal olfactory cortex differ by projecting to the Ammons horn and *via* corona radiata respectively, which might imply that he included the present area 28L in his spheno-occipital ganglion. However, since Cajal (1955) was not concerned with mapping the border on the basis of efferent connections, we may assume with Scalia (1966) that he defined the spheno-occipital ganglion by its structure (cp. Cajal 1955, p. 176). It is consistent with this that Cajal (1955, p. 30, man, p. 32, guinea pig and rabbit) described cell clusters and alternate slight condensations and rarefactions in layer II of *Region olfactive principale,* a feature more characteristic of the ventrolateral entorhinal area than of the pyriform cortex (see also Pigache, 1970, pp. 36, 38; Stephan, 1975, pp. 430, 649)

Brodman (1909) introduced the name area entorhinalis (i. e., inside the rhinal fissure). Although Brodman (1909) did not describe or illustrate the transition between the entorhinal area and the prepyriform cortex, it appears from his diagrams that he defined the former in the manner subsequently illustrated by M. Rose (see below). Brodman (1909) also subdivided it, in lower mammals, into a part nearer the rhinal fissure and a part nearer the parasubiculum (Brodman, 1909, Fig. 103: Pteropus; Fig. 105: Cercoleptes; Fig. 107: Lepus; Fig. 111: Erinaceus), thus suggesting the current subdivision into medial and lateral parts. M. Rose (1912) first accepted Brodman's principle of subdivision into medial and lateral parts for several other small mammals, including mouse and guinea pig (M. Rose, 1912, Figs. 31, 35) although the accompanying documentation of fibre architectonics (his Figs. 10, 15) and cytoarchitectonics (his Tafel 3, 4, 5, 11) is inconsistent in the identification of the two subareas. Later, he proposed much more detailed parcellations of the entorhinal area for various mammals (1926), mouse (1929), rabbit (1931) and primates (1927). Thus, in mouse (1929, Abb. 4 and Tafel 14−17) he introduced three rostroventral fields (e1−e3) keeping a subdivision into a medial and lateral subarea (e5−e4) dorsally. These subareas bear no unequivocal relation to the subdivisions proposed here. However, roughly speaking e5 corresponds to area 28M; e4 and a posterior part of e1 corresponds to 28L; the remaining part of e1 together with e2 corresponds to area 28L' and an adjacent bit of pyriform cortex; and finally the medial part of e3 corresponds to area TR. Rose did not illustrate anything precisely like the present area 28M'. In the rabbit (1931, Abb. 4 and Tafel 12−19) he felt compelled to define nine subareas to do justice to the gradual mediolateral and caudorostral changes seen in Nissl material.

Popoff and Popoff (1929), following Rose, presented a very detailed map of entorhinal subareas in the rat, equally difficult to compare with the present subdivisions. Their areas *ea* and *eb* correspond partly to areas 28M and 28L. The bands *x, y* and *ye,* situated dorsal to *ea* may lie partly just dorsal to the present area 28, partly just within its dorsal border and partly coincide with the parasubiculum. Further rostroventrally *ed1* and the ventral end of *ea* could correspond in part to area 28M'. The subdivisions *ec, ee* and *ef1−ef2* are Popoff and Popoff's attempt at diagramizing the gradual changes occurring between area 28L and the pyriform cortex, while their *ed2* and *ed3* cover approximately the present area TR.

These detailed subdivisions based merely on cytoarchitectonics (Rose, Popoff and Popoff, loc. cit.) were criticized by Lorente de Nó (1934, pp. 158−165). Based on a study of Golgi material, the latter author tentatively distinguished the following three parts within the entorhinal area: *Part A,* lacking layer IIIa and showing other characteristics of the present area 28L, *Part C,* apparently corresponding quite well with the present area 28M. *Part B,* corresponding more or less to the transition between the present areas 28M and 28L (see Lorente de Nó, 1934, Figs. 2, 2a). He further asserted (Lorente de Nó, 1934, p. 161) that any division into subareas of the entorhinal area should be made along its dorsoventral (caudorostral) axis and that this fundamental organization is constant through a phylogenetic series (mouse, rabbit, cat and monkey). In his opinion, the mediolateral differences in Golgi-architecture were quite gradual, so that he felt compelled to subdivide his subareas B and C into 4 or more sub-subareas. Dorsoventral changes were not denied, but held to be less fundamental than the mediolateral ones and not to support the definition of subareas. In particular he argued that the plan of connections does not change as one goes along the dorsoventral axis of the area. However, Lorente de Nó's (1934) views on entorhinal connections, based on Golgi-studies, have not been wholly confirmed by recent experimental studies.

Thus, he stated that the lateral part gives off the perforant path, the medial part gives off the alvear path and the intermediate part gives rise to both types of fibres. Although his argument might be reformulated according to recent studies of entorhinal connections (Hjorth-Simonsen, 1972; Hjorth-Simonsen and Jeune, 1972; Van Hoesen et al., 1972, Van Hoesen and Pandya, 1975b), putting medial perforant path in place of alvear path and lateral perforant path in place of just perforant path, it remains to confirm that these connections characterize also area 28M' versus area 28L'.

Several investigators, accepting Lorente de Nó's (1934) emphasis on the more fundamental nature of mediolateral than of caudorostral changes within area 28, proceeded to deal with horizontal sections at rather dorsal levels showing mainly areas 28M and 28L.

Blackstad (1956) gave a detailed cyto- and fibrearchitectonic description of the medial and lateral parts in the rat and showed that the commissural afferent pattern corroborated this subdivision. He described a transition rather than a precise border between the medial and lateral part both in Nissl stained and silver impregnated material, (Blackstad, 1956, pp. 428, 430), but remarked (p. 430) that an accurate limit may nevertheless often be seen. Later histochemical studies (Storm-Mathisen and Blackstad, 1964; Mellgren and Blackstad, 1967; Geneser-Jensen and Blackstad, 1971) and the definition of both a medial and a lateral perforant path (Hjorth-Simonsen, 1972; Hjorth-Simonsen and Jeune, 1972; Van Hoesen et al., 1972) further confirmed the difference between areas 28M and 28L. However, a transitional or intermediate zone between the two was again postulated (Hjorth-Simonsen, 1972; Van Hoesen et al., 1972).

Regarding the ventral parts of the entorhinal area, Geneser Jensen and Blackstad (1971) described a modified cholinesterase pattern in this region without attempting to delimit and identify the modified parts systematically. They reported a sharp entorhinal-pyriform border. Meanwhile, investigators primarily interested in the termination of olfactory afferents were particularly concerned with the ventral parts of the entorhinal area. Thus, Scalia (1966) accepted Rose's (1931) nine subdivisions of the rabbit entorhinal area, equating that author's e8 and e9 with Cajal's spheno-occipital ganglion. Dennis and Kerr (1968) and Kerr and Dennis (1972) proposed the designation parahippocampal (PH) cortex for all cortex between the pyriform cortex and Cajal's spheno-occipital ganglion, but gave no detailed morphological analysis of it. Price and Powell (1971, rat) emphasized the gradual transition of cytoarchitectonic characteristics between the pyriform cortex and the spheno-occipital ganglion of Cajal. They defined an entorhinal-pyriform transition area characterized by: Lack of a deep plexiform layer (i. e. present IIIa); split layer II with island formation in the outer sublayer; secondary olfactory afferents to the molecular layer. In a later paper Price (1973) reaffirmed the presence of gradual changes between the pyriform and entorhinal cortices, but preferred to identify most of the transitional zone as area 28L in the sense of Blackstad (1956).

In a recent important series of papers on entorhinal connections in the monkey Van Hoesen et al. (Van Hoesen and Pandya, 1975a,b; Van Hoesen et al., 1972, 1975) divided the entorhinal area of this species into area 28a (medial part), area 28i (intermediate part) and area 28b (lateral part) on the basis of cytoarchitecture and connections. In addition, the lateralmost part of the entorhinal area, situated in the entorhinal bank of the rhinal fissure was set apart as prorhinal area. These authors made no provision for separate ventral parts and drew a definite border between the entorhinal area and what they called the periamygdaloid cortex (would be part of the present pyriform cortex, see Nomenclature). Thus, although they recognized caudorostral as well as mediolateral changes in structure, they agreed with Lorente de Nó (1934) in attaching more importance to the latter changes. The prorhinal area they divided in a small, rostrally situated, subarea 1 and an elongated subarea 2, bordering prorhinal area 1 and area 28a, 28i and 28b laterally. Prorhinal area 1 was defined as resembling Price and Powell's (1971) entorhinal-pyriform transition area. However, this definition would seem to be invalidated by the fact that Price (1973) considered the latter area as identical to area 28L of Blackstad (1956), which again would be homologous to area 28b of Van Hoesen et al. (loc. cit.).

The above selective review emphasizes studies in lower mammals, most relevant to the findings reported here. Stephan (1975) should be consulted for a complete survey of the literature through 1974.

References

Allison, A. C.: The secondary olfactory areas in the human brain. J. Anat. (Lond.) 88, 481–488 (1954)

Blackstad, T. W.: Commissural connections of the hippocampal region in the rat, with special reference to their mode of termination. J. comp. Neurol. 105, 417–538 (1956)

Blackstad, T. W., Fuxe, K., Hökfeldt, T.: Noradrenaline nerve terminals in the hippocampal region of the rat and the guinea pig. Z. Zellforsch. 78, 463–473 (1967)

Brodmann, K.: Vergleichende Lokalisationslehre der Grosshirnrinde in ihren Prinzipien dargestellt auf Grund des Zellenbaues. 324 pp. Leipzig: J. A. Barth 1909

Cajal, S. Ramón y: Estudios sobre la corteza cerebral humana. IV Estructura de la corteza cerebral olfativa del hombre y maniferos. Trab. Lab. Invest. Biol. Madr. 1, 1–140 and further papers in the same volume, pp. 141–150, 159–188, 189–206 (1901–1902)

Cajal, S. Ramón y: Histologie du système nerveux de l'homme et des vertébrés. T. II. Paris: A. Maloine 1911; reprinted Madrid: Instituto Cajal 1955

Cajal, S. Ramón y: Studies on the cerebral cortex (limbic structures), trans. L. M. Kraft. English translation of Cajal (1901–1902). 179 pp. London: Lloyd-Luke 1955

Dennis, B. J., Kerr, D. I. B.: An evoked potential study of centripetal and centrifugal connections of the olfactory bulb in the cat. Brain Res. 11, 373–396 (1968)

Domesick, V. B.: Projections from the cingulate cortex in the rat. Brain Res. 12, 296–320 (1969)

Domesick, V. B.: The fasciculus cinguli in the rat. Brain Res. 20, 19–32 (1970)

Domesick, V. B.: Thalamic projections in the cingulum bundle to the parahippocampal cortex of the rat. Anat. Rec. 175, 308 (1973)

Geneser-Jensen, F. A., Blackstad, T. W.: Distribution of acetyl cholinesterase in the hippocampal region of the guinea pig. I. Entorhinal area, parasubiculum, and presubiculum. Z. Zellforsch. **114**, 460–481 (1971)

Geneser-Jensen, F. A., Haug, F.-M. Š., Danscher, G.: Distribution of heavy metals in the hippo-campal region of the guinea pig. A light microscope study with Timm's sulphide silver method. Z. Zellforsch. **147**, 441–478 (1974)

Hall, E., Geneser-Jensen, F. A.: Distribution of acetylcholinesterase and monoamine oxidase in the amygdala of the guinea pig. Z. Zellforsch. **120**, 204–221 (1971)

Haug, F.-M. Š.: Heavy metals in the brain. A light microscope study of the rat with Timm's sulphide silver method. Methodological considerations and cytological and regional staining patterns. Advances in Anatomy, Embryology and Cell Biology 47, 1–71 (1973)

Haug, F.-M. Š.: Light microscopical mapping of the hippocampal region, the pyriform cortex and the corticomedial amygdaloid nuclei of the rat with Timm's sulphide silver method. I. Area dentata, hippocampus and subiculum. Z. Anat. Entwickl.-Gesch. **145**, 1–27 (1974)

Haug, F.-M. Š.: On the normal histochemistry of trace metals in the brain. J. Hirnforsch. **16**, 146–158 (1975)

Haug, F.-M. Š.: Laminar distribution of afferents in the allocortex, visualized with Timm's sulphide silver method for "heavy" metals. Abstract, 7th International Neurobiology Meeting, Göttingen, Sept. 15–19, 1975 (In press, 1976)

Heimer, L.: Synaptic distribution of centripetal and centrifugal nerve fibres in the olfactory system of the rat. An experimental anatomical study. J. Anat. (Lond.) **103**, 413–432 (1968)

Hjorth-Simonsen, A.: Hippocampal efferents to the ipsilateral entorhinal area: An experimental study in the rat. J. comp. Neurol. **142**, 417–438 (1971)

Hjorth-Simonsen, A.: Projection of the lateral part of the entorhinal area to the hippocampus and fascia dentata. J. comp. Neurol. **146**, 219–232 (1972)

Hjorth-Simonsen, A.: Some intrinsic connections of the hippocampus in the rat: An experimental analysis, J. comp. Neurol. **147**, 145–162 (1973).

Hjorth-Simonsen, A., Jeune, B.: Origin and termination of the hippocampal perforant path in the rat studied by silver impregnation. J. comp. Neurol. **144**, 215–232 (1972)

Katz, L., Levinthal, C.: Interactive computer graphics and representation of complex biological structures. Ann. rev. Biophys. Bioeng. **1**, 465–504 (1972)

Kerr, D. I. B., Dennis, B. J.: Collateral projection of the lateral olfactory tract to entorhinal cortical areas in the cat. Brain Res. **36**, 399–403 (1972)

Krettek, J. E., Price, J. L.: Projections from the amygdala to the perirhinal and entorhinal cortices and the subiculum. Brain. Res. **71**, 150–154 (1974)

Krieg, W. J. S.: Connections of the cerebral cortex. I. The albino rat. A. Topography of the cortical areas. J. comp. Neurol. **84**, 221–275 (1946)

König, J. F. R., Klippel, R. A.: The rat brain. A stereotaxic atlas of the forebrain and lower parts of the brain stem. Baltimore: Williams and Wilkins 1963

Llinás, R., Hillman, D. E.: A multipurpose tridimensional reconstruction computer system for Neuroanatomy. In: Golgi Centennial Symposium: Perspectives in Neurobiology (M. Santini, ed.), pp. 71–79. New York: Raven Press 1975

Loo, Y. T.: The forebrain of the opossum, Didelphys Virginiana. Part II. Histology. J. comp. Neurol. **52**, 1–148 (1931)

Lorente de Nó, R.: Studies on the structure of the cerebral cortex. I. The area entorhinalis. J. Psychol. Neurol. (Lpz.) **45**, 381–438 (1933)

Lorente de Nó, R.: Studies on the structure of the cerebral cortex. II. Continuation of the study of the ammonic system. J. Psychol. Neurol. (Lpz.) **46**, 113–177 (1934)

Mellgren, S. I.: Distribution of succinate-, α-Glycerophosphate-, NADH-, and NADPH dehydroge-nases (tetrazolium reductases) in the hippocampal region of the rat during postnatal develop-ment. Z. Zellforsch. **141**, 347–373 (1973a)

Mellgren, S. I.: Distribution of acetylcholinesterase in the hippocampal region of the rat during postnatal development. Z. Zellforsch. **141**, 375–400 (1973b)

Mellgren, S. I., Blackstad, T. W.: Oxidative enzymes (tetrazolium reductases) in the hippocampal region of the rat. Distribution and relation to architectonics. Z. Zellforsch. **78**, 167–207 (1967)

Mellgren, S. I., Geneser-Jensen, F. A.: Distribution of monoamine oxidase in the hippocampal region of the rat. Z. Zellforsch. **124**, 354–366 (1972)

Meyer, M., Allison, A. C.: An experimental investigation of the connexions of the olfactory tracts in the monkey. J. Neurol. Neurosurg. Psychiat. **12**, 274–286 (1949)

Pigache, R. M.: The anatomy of "paleocortex". A critical review. Ergebn. Anat. Entwickl.-Gesch. **43**, 1–62 (1970)

Popoff, I., Popoff, N.: Allocortex bei der Ratte (mus decumanus). J. Psychol. Neurol. (Lpz.) **39**, 257–322 and Tafel 22–33 (1929)

Price, J. L.: An autoradiographic study of complementary laminar patterns of termination of afferent fibers to the olfactory cortex. J. comp. Neurol. **150**, 87–108 (1973)

Price, J. L., Powell, T. P. S.: Certain observations on the olfactory pathway. J. Anat. (Lond.) **110**, 105–126 (1971)

Rakic, P., Stensaas, L. J., Sayre, E. P., Sidman, R. L.: Computeraided three-dimensional reconstruction and quantitative analysis of cells from serial electron microscopic montages of foetal monkey brain. Nature (Lond.) **250**, 31–34 (1974)

Rose, M.: Histologische Lokalisation der Grosshirnrinde bei kleinen Säugetieren (Rodentia, Insectivora, Chiroptera). J. Psychol. Neurol. (Lpz.), 19 (Erg. b. 2): 391–479 and Tafel 1–15 (1912)

Rose, M.: Der Allocortex bei Tier und Mensch. I. Teil. J. Psychol. Neurol. (Lpz.) **34**, Heft 1 u. 2, 1–111 and Tafel 1–30 (1926)

Rose, M.: Die sog, Riechrinde beim Menschen und beim Affen. II. Teil des "Allocortex bei Tier und Mensch". J. Psychol. Neurol. (Lpz.) **34**, Heft 6, 262–401 and Tafel 42–76 (1927)

Rose, M.: Cytoarchitektonischer Atlas der Grosshirnrinde der Maus. J. Psychol. Neurol. (Lpz.) **40**, 1–51 and Tafel 1–29 (1929)

Rose, M.: Cytoarchitektonischer Atlas der Grosshirnrinde des Kaninchens. J. Psychol. Neurol. (Lpz.) **43**, 353–440 and Tafel 1–43 (1931)

Rose, J. E., Woolsey, C. N.: Structure and relations of limbic cortex and anterior thalamic nuclei in rabbit and cat. J. comp. Neurol. 89, 279–347 (1948)

Scalia, F.: Some olfactory pathways in the rabbit brain. J. comp. Neurol. **126**, 285–310 (1966)

Scalia, F., Winans, S. S.: The differential projections of the olfactory bulb and accessory olfactory bulb in mammals. J. comp. Neurol. **161**, 31–56 (1975)

Segal, M., Landis, S. C.: Afferents to the hippocampus of the rat studied with the method of retrograde transport of horseradish peroxidase. Brain Res. **78**, 1–15 (1974)

Shipley, M. T.: The topographical and laminar organization of the presubiculum's projection to the ipsi- and contralateral entorhinal cortex in the guinea pig. J. comp. Neurol. **160**, 127–146 (1975)

Shipley, M. T., Geneser-Jensen, F. A., Meier, A.: Correlated histochemical and experimental evidence for a subdivision of the entorhinal area of the guinea pig. Cell Tiss. Res. **150**, 455–462 (1974)

Shipley, M. T., Sørensen, K. E.: On the laminar organization of the anterior thalamus projections to the presubiculum in the guinea pig. Brain Res. **86**, 473–477 (1975)

Smith, R. W., White, L. E., jr.: The fiberarchitectonics of the cat hippocampal formation. J. comp. Neurol. **123**, 11–28 (1964)

Stephan, H.: Allocortex. Handbuch der mikroskopischen Anatomie des Menschen (W. Bargman, ed.) Vol. 4(Nervensystem), Part 9. Berlin-Heidelberg-New York: Springer 1975

Storm-Mathisen, J., Blackstad, T. W.: Cholinesterase in the hippocampal region. Distribution and relation to architectonics and afferent systems. Acta anat. (Basel) **56**, 216–253 (1964)

Timm, F.: Zur Histochemie der Schwermetalle. Das Sulfid-Silberverfahren. Dtsch. Z.ges.gerichtl. Med. **46**, 706–711 (1958a)

Timm, F.: Zur Histochemie des Ammonshorngebietes. Z. Zellforsch. **48**, 548–555 (1958b)

Van Hoesen, G. W., Pandya, D. N.: Some connections of the entorhinal (area 28) and perirhinal (area 35) cortices of the rhesus monkey. I. Temporal lobe afferents. Brain Res. **95**, 1–24 (1975a)

Van Hoesen, G.W., Pandya, D. N.: Some connections of the entorhinal (area 28) and perirhinal (area 35) cortices of the rhesus monkey. III. Efferent connections. Brain Res. **95**, 35–59 (1975b)

Van Hoesen, G. W., Pandya, D. N., Butters, N.: Cortical afferents to the entorhinal cortex of the rhesus monkey. Science **175**, 1471–1473 (1972)

Van Hoesen, G. W., Pandya, D. N., Butters, N.: Some connections of the entorhinal (area 28) and perirhinal (area 35) cortices of the rhesus monkey. II. Frontal lobe afferents. Brain Res. 95, 25–38 (1975)

Vaz Ferreira, A.: The cortical areas of the albino rat studied by silver impregnation. J. comp. Neurol., 95, 177–243 (1951)

Voigt, G. E.: Histologische Versilberungen. Habil.-Schr. Jena (1951)

White, L. E., jr.: Ipsilateral afferents to the hippocampal formation in the albino rat. I. Cingulum projections. J. comp. Neurol. 113, 1–42 (1959)

White, L. E., jr.: Olfactory bulb projections of the rat. Anat. Rec. 152, 465–480 (1965)

Zunino, G.: Die Myeloarchitektonische Differenzierung der Grosshirnrinde beim Kaninchen (Lepus cuniculus). J. Psychol. Neurol. (Lpz.) 14, 38–70 (1909)

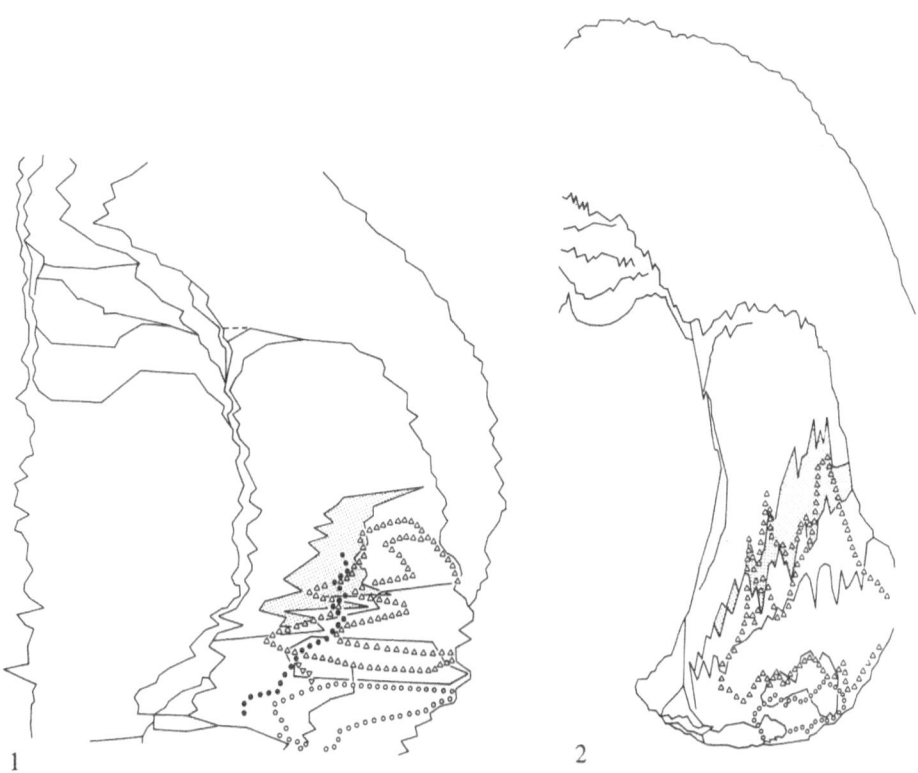

1   2

Figs. 1 and 2. Posterior projections of parahippocampal areas in the rat, reconstructed graphically as described in the text. Labels and level numbers have been omitted. Symbols as in Figs. 3–10. Compare with Fig. 4 and 9, respectively, to get an impression of how much the original curves have been smoothed in the following Figs. 3–10

Figs. 3–10. Surface maps of the hippocampal region and pyriform cortex, reconstructed as described in the text from the right hemispheres of three rat brains. The surface maps are based on the three series of sections from which Figs. 11–48 are taken. To refer from a photograph to the corresponding surface maps, find level number in upper left or right hand cornet of the photograph. Then locate it on the scales of level numbers in each map and place a ruler across the map in this position. Figs. 3–5, 6–8 and 9–10 represent the horizontal, frontal and sagittal series, illustrated in Figs. 11–28, 29–39 and 40–48, respectively. The same symbols and abbreviations are used throughout (see: "Abbreviations and symbols for all figures"). In the reconstructions, thick lines indicate contours or fissures of the brain, alternate dots and dashes indicate cut faces of the brain. Thin lines and filled symbols indicate interareal borders in the sulphide silver pattern and open symbols indicate cytoarchitectonic borders. The presence of gradual changes rather than sharp interareal borders is shown either by giving approximate limits of the transition zone (cytoarchitectonic transition in layer II between areas *28M* and *28L*) or by stippling the entire transitional zone (change of Timm staining in layer II between areas *28M* and *28L*, gradual changes between areas *28L* and the pyriform cortex with both methods). Details of the pyriform cortex and the amygdala included in the present reconstructions will be described elsewhere

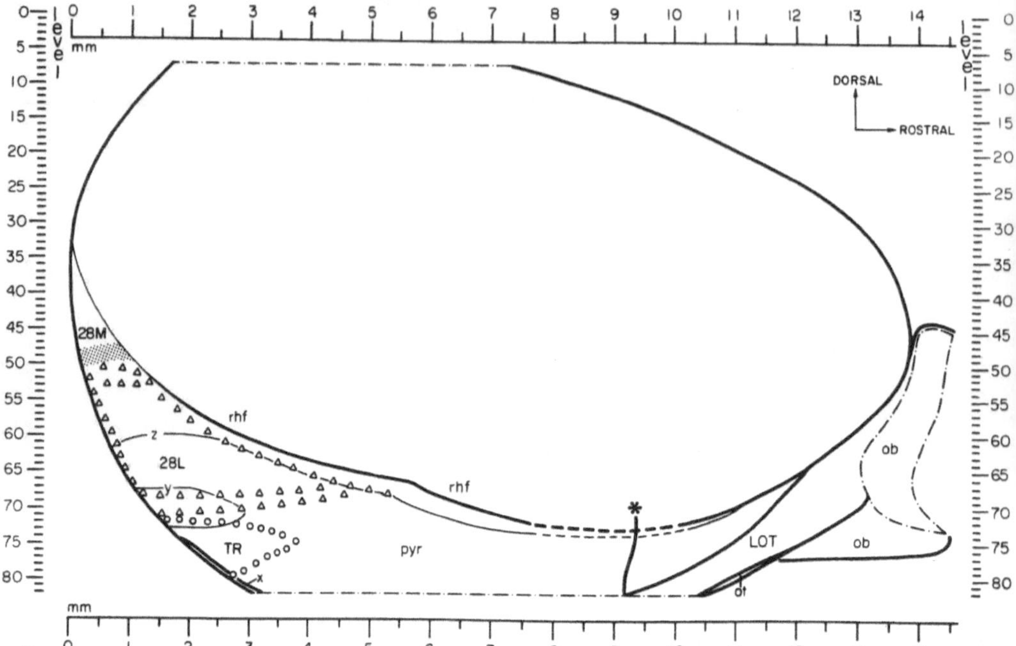

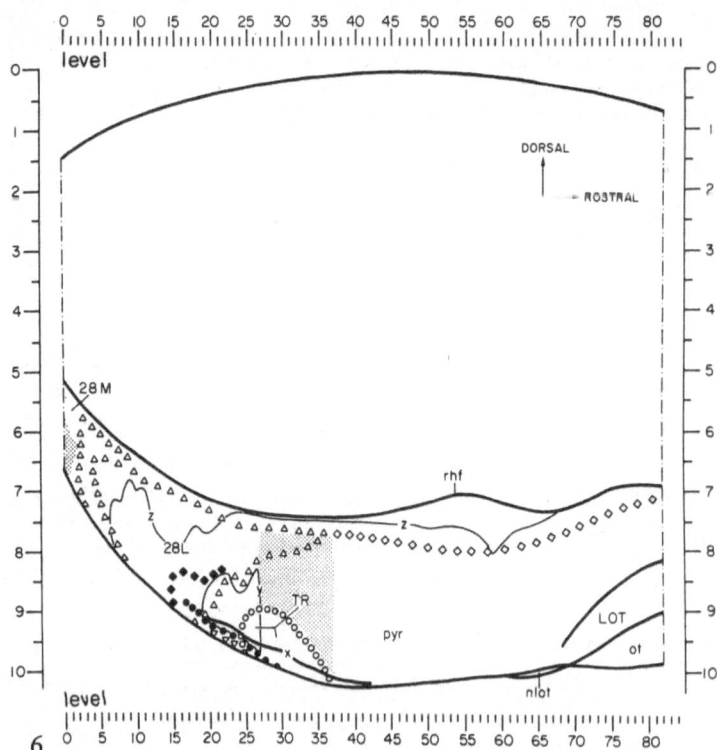

Figs. 3 and 6

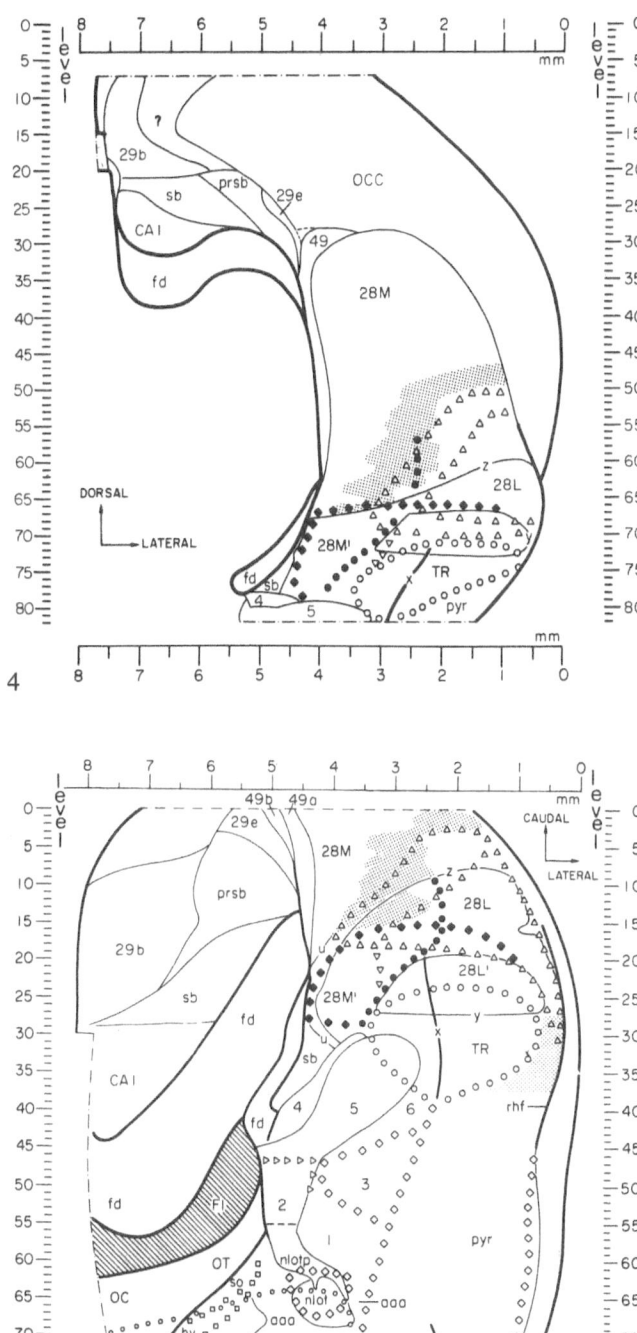

Figs. 4 and 7

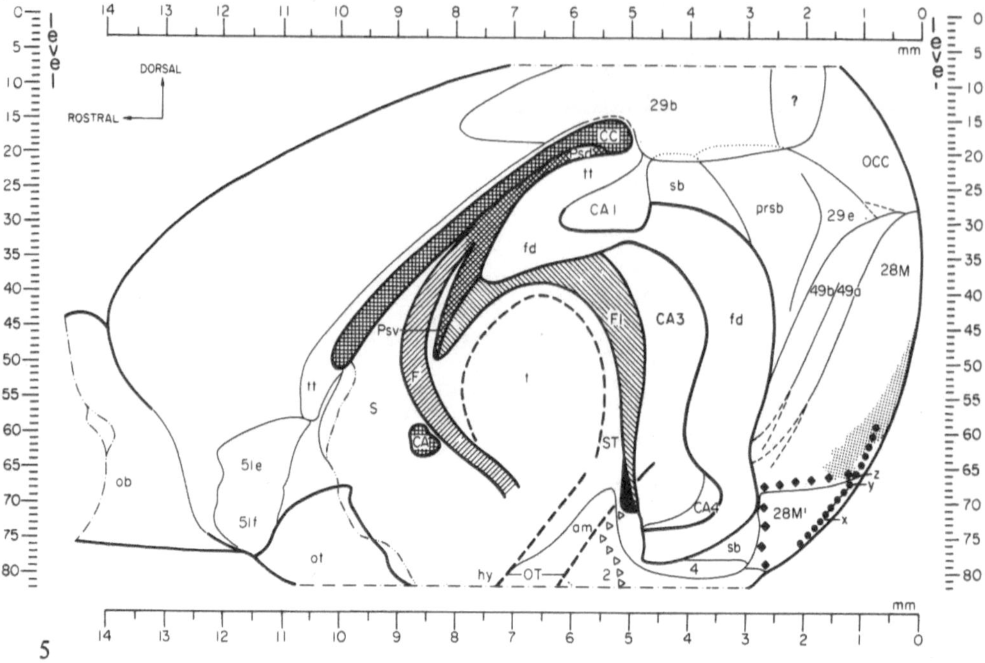

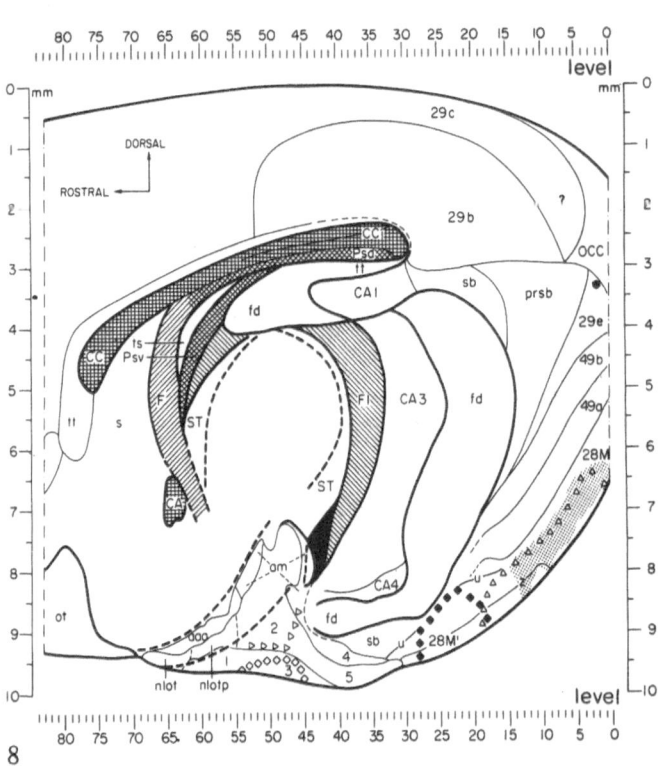

Figs. 5 and 8

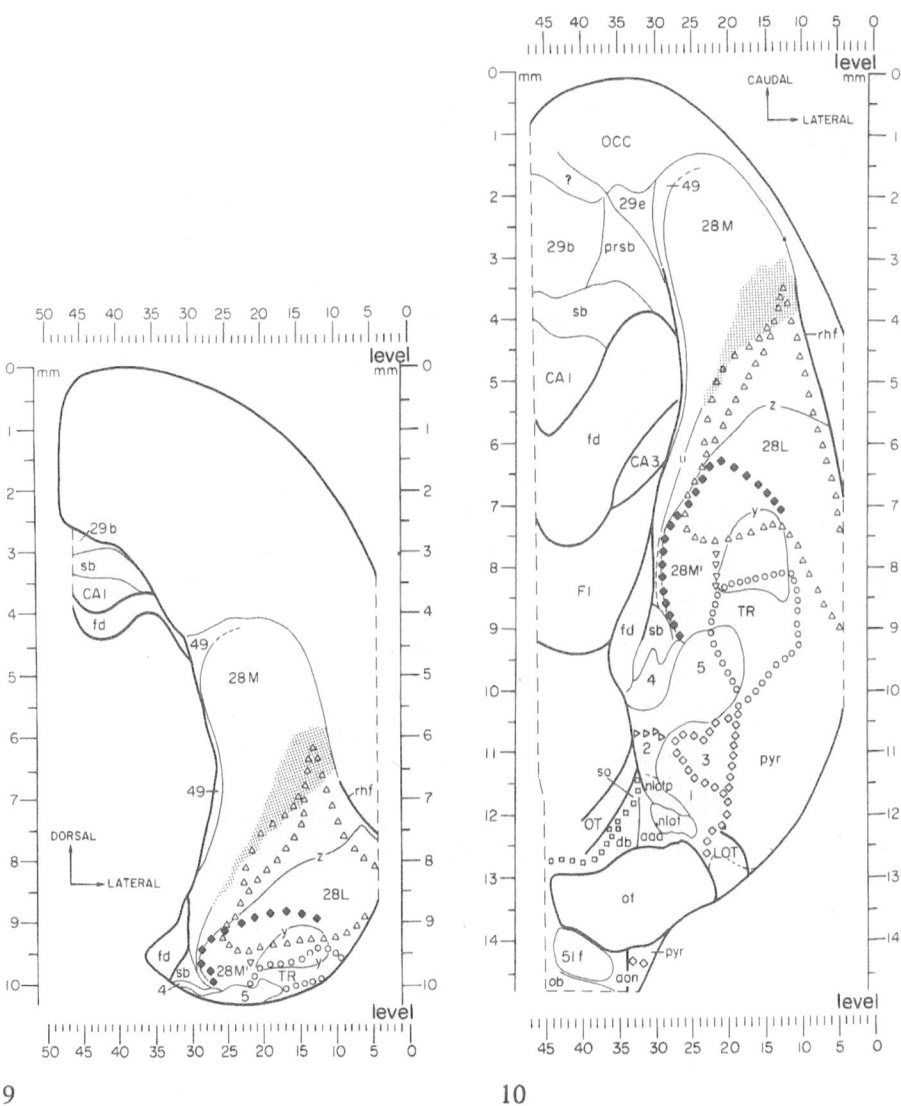

9

10

Figs. 9 and 10

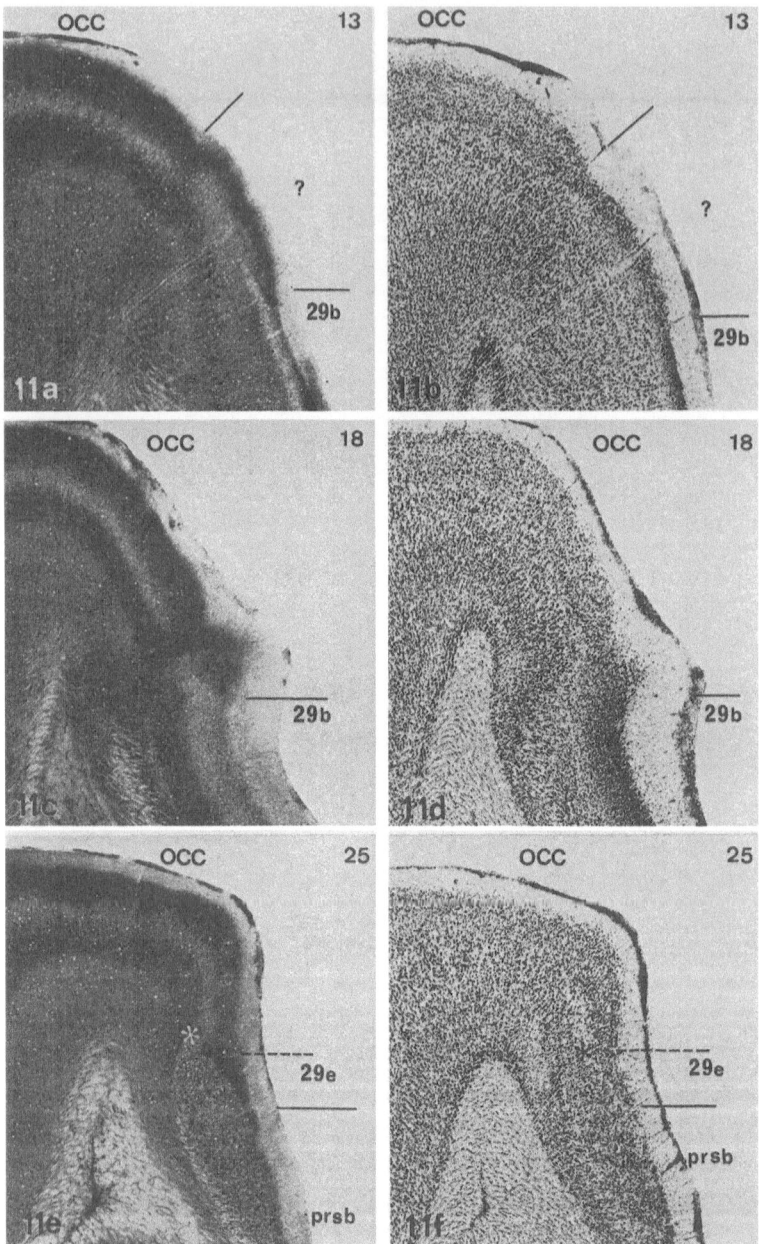

Fig. 11a–f. Horizontal series. × 18
General explanation for all photomicrographs (Figs. 11–48): See footnote 4, p. 12

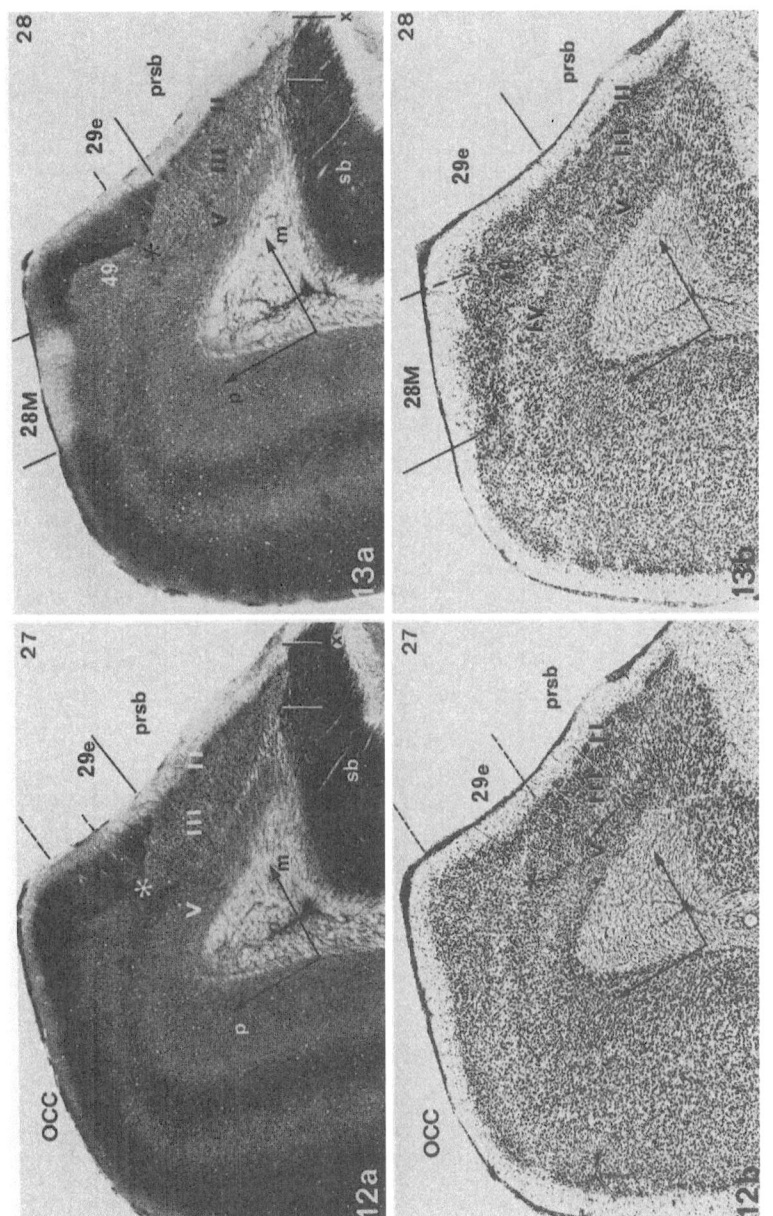

Figs. 12a and 13. Horizontal series. × 18

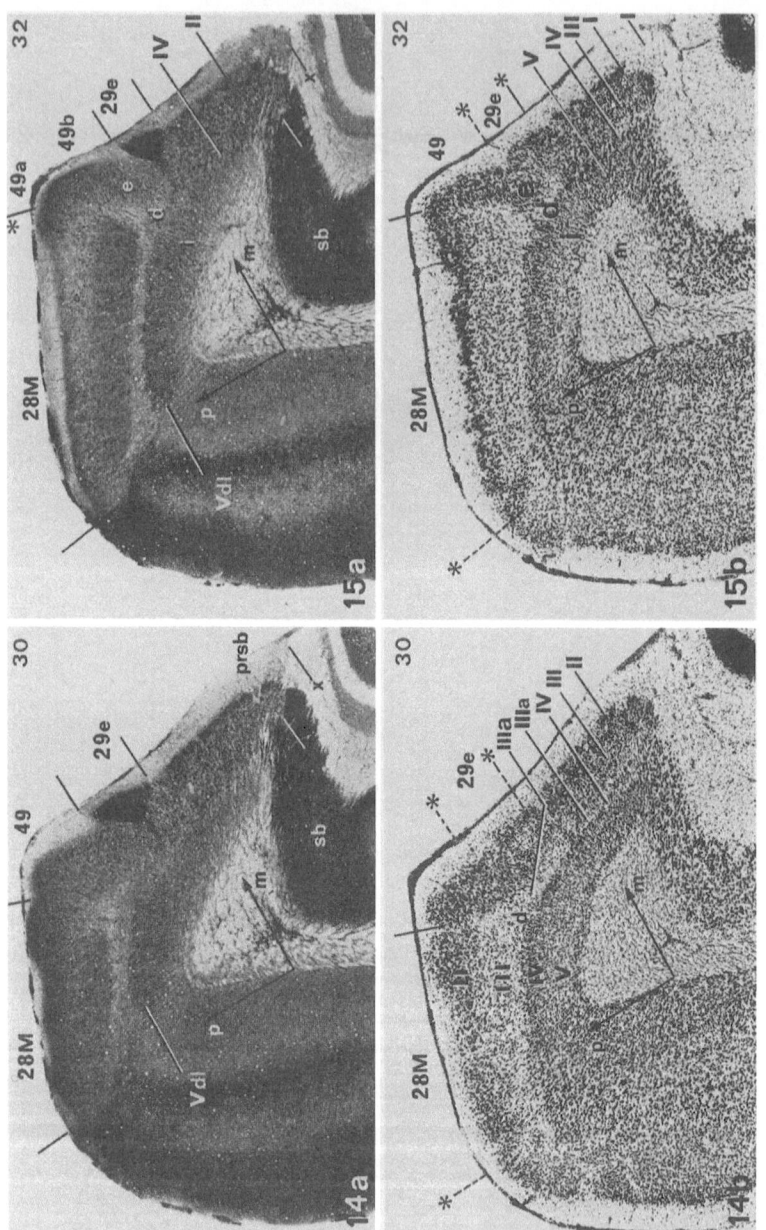

Figs. 14 and 15. Horizontal series. Note that the sulphide silver pattern more than the cytoarchitectonics suggests a sharp dorsolateral border to area 28M. × 18

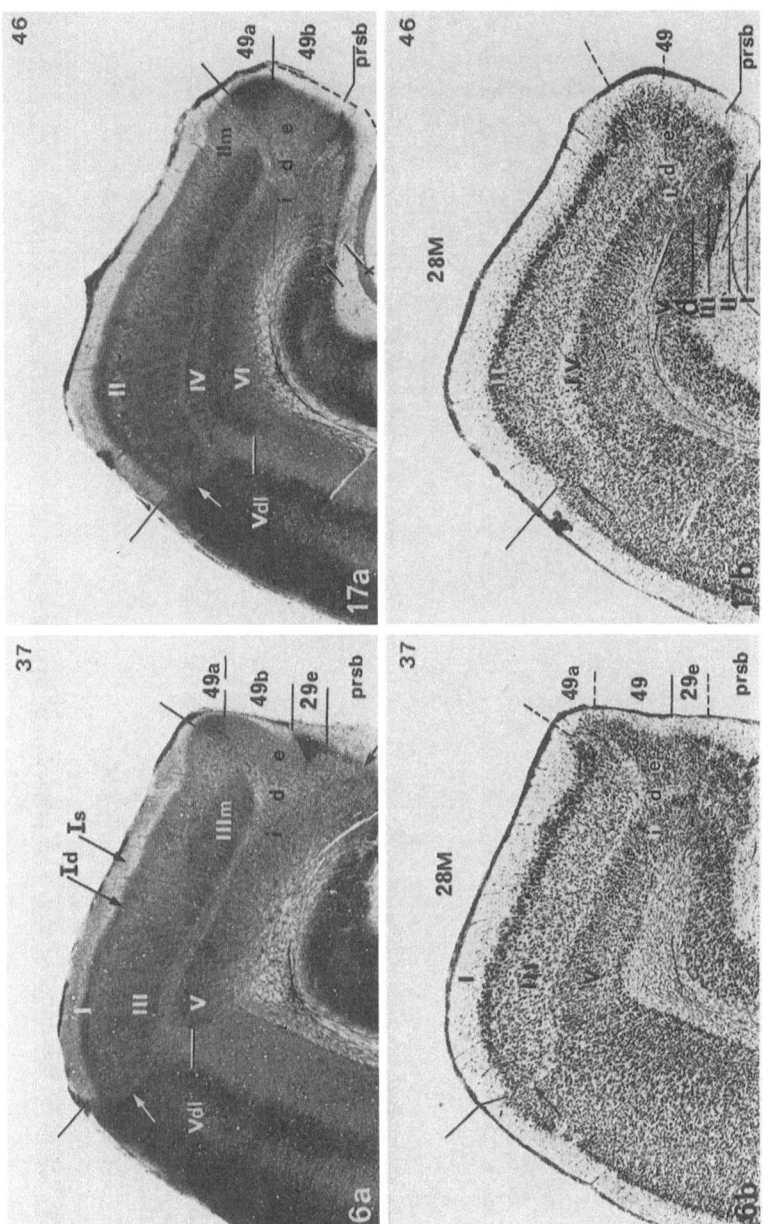

Figs. 16 and 17. Horizontal series. × 18

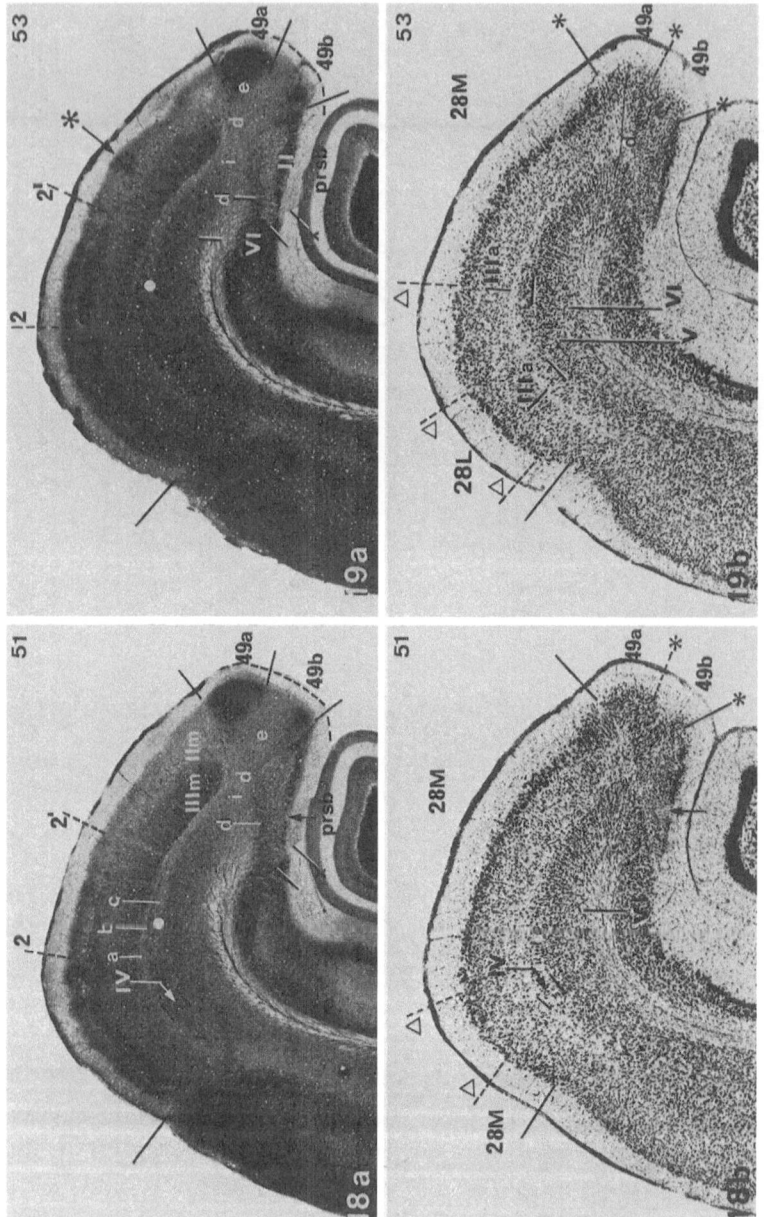

Figs. 18 and 19. Horizontal series. a–b–c: sublaminae of layer IV. × 18

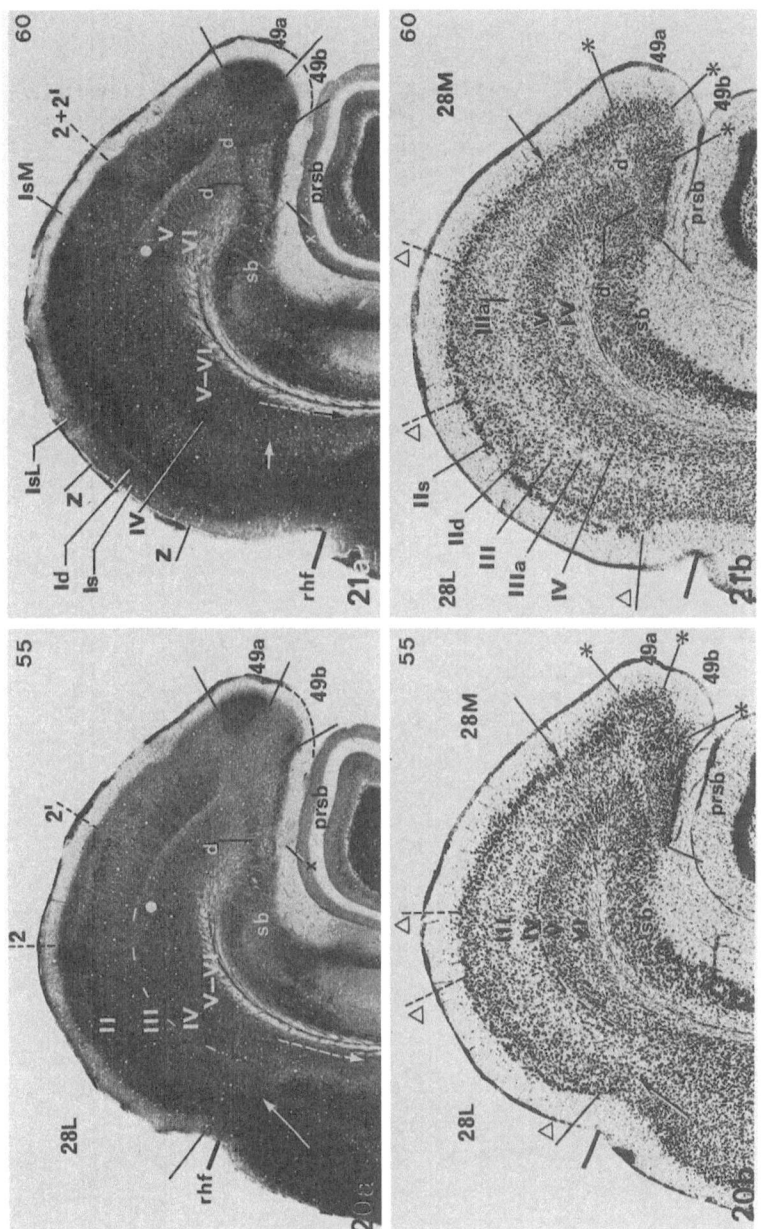

Figs. 20 and 21. Horizontal series. Note medial end of layer II becoming looser and less characteristic of area *28M* as we go ventral-wards (medial to arrows in 20b and 21b). Compare with preceding and succeeding figures. × 18

51

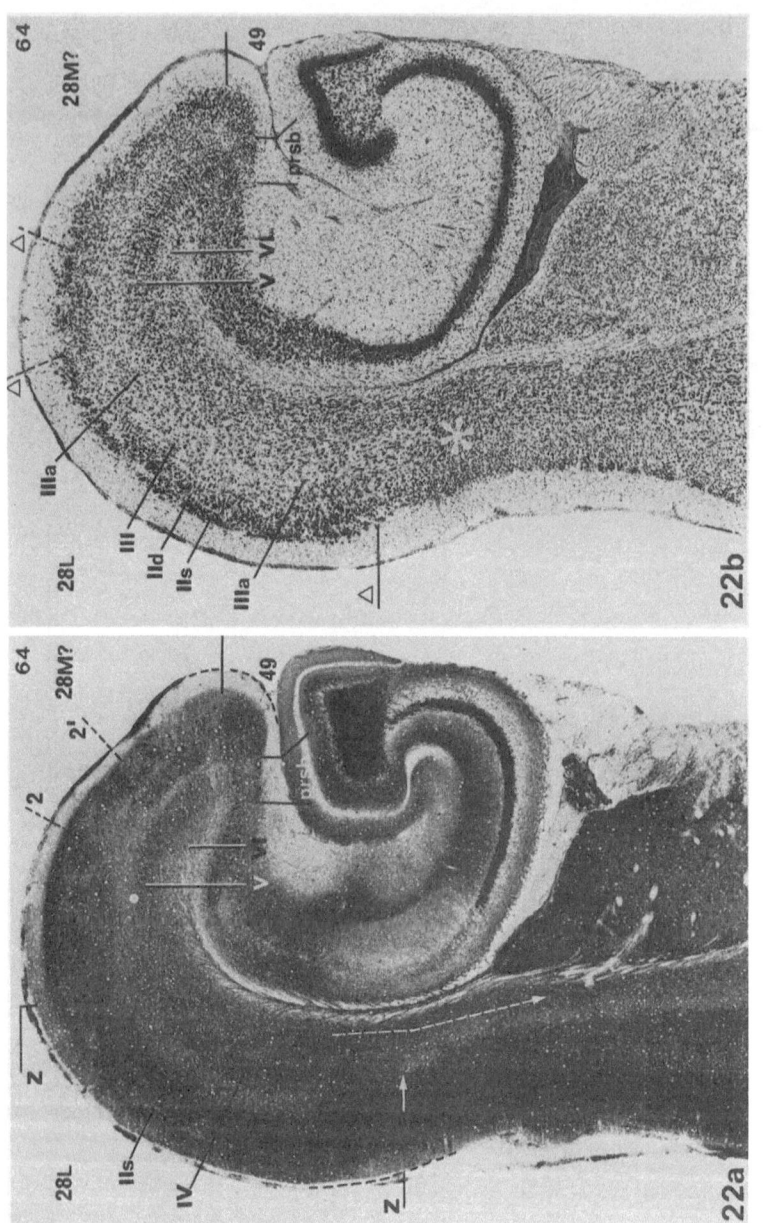

Fig. 22a and b. Horizontal series. Compression artefact at asterisk. × 18

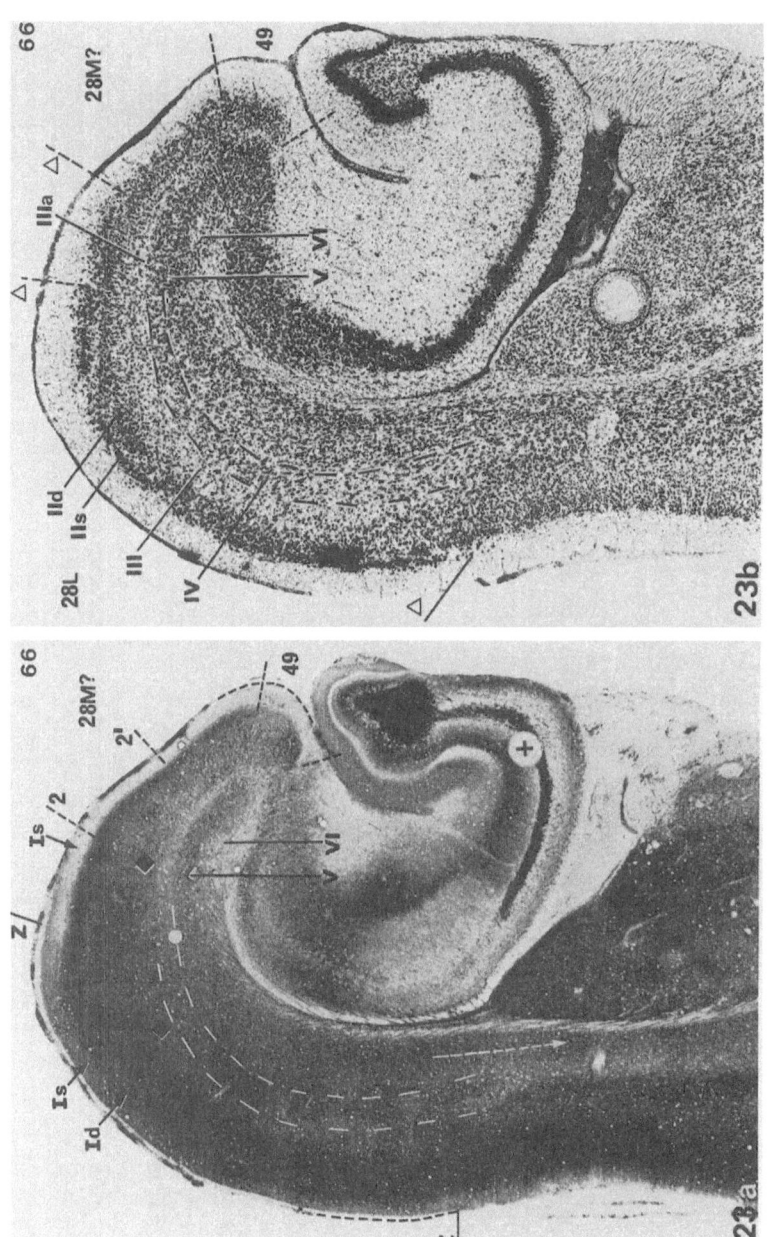

Fig. 23a and b. Horizontal series. × 18

53

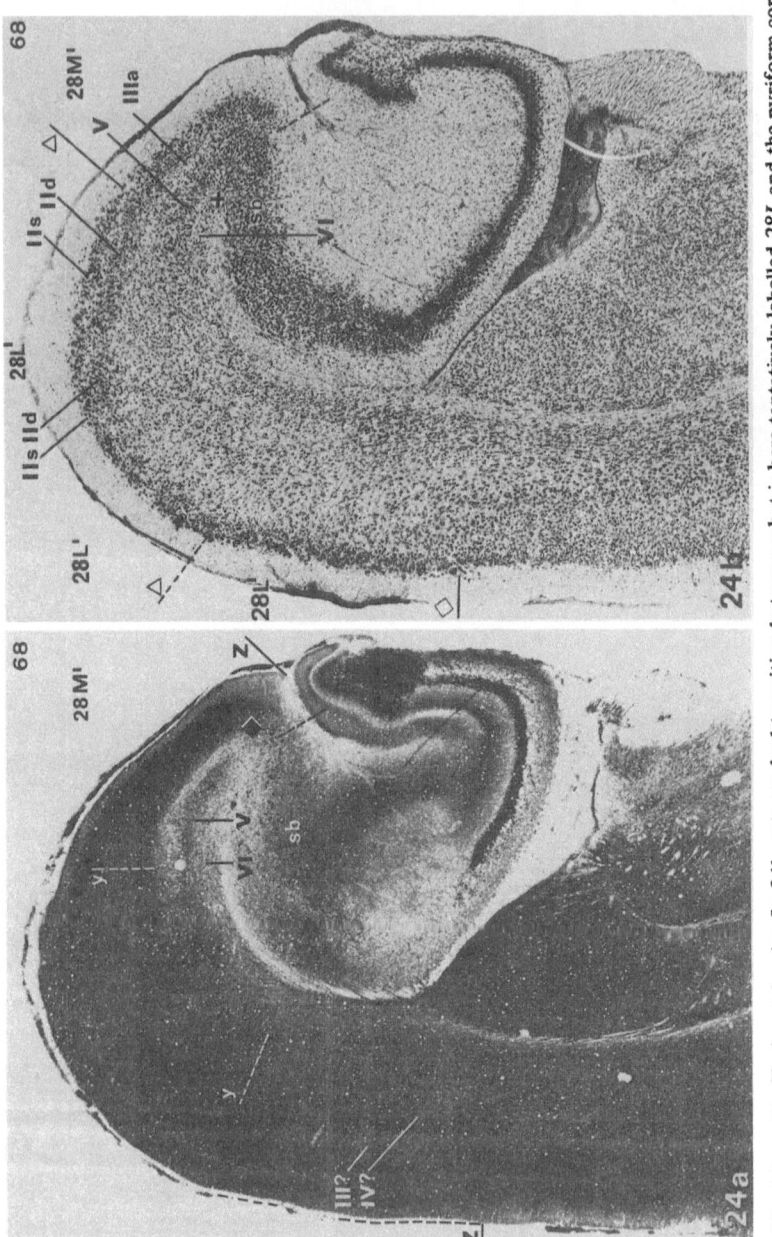

Fig. 24a and b. Horizontal series. In 24b note gradual transition between what is here tentatively labelled *28L* and the pyriform cortex, both with regard to layer II and when considering in addition the deeper layer (compare with Figs. 23b, 25b and 26b). × 18

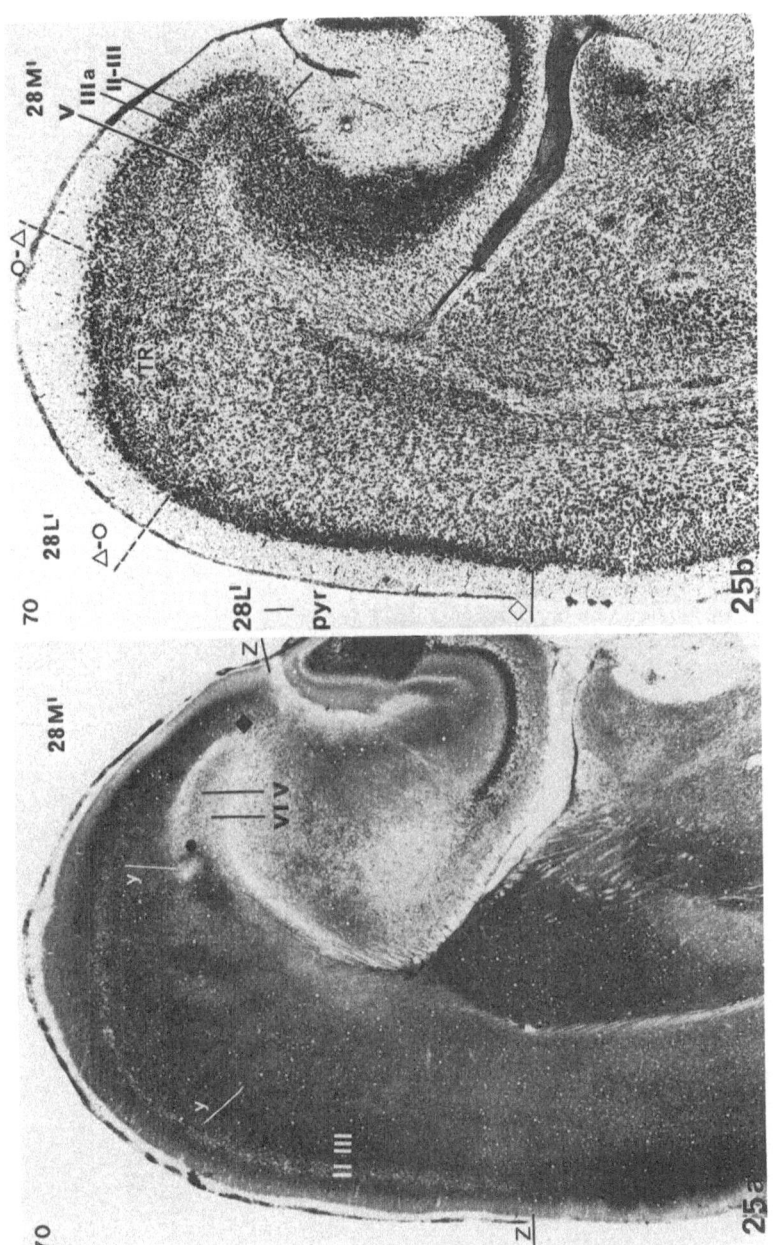

Fig. 25a and b. Horizontal series. × 18

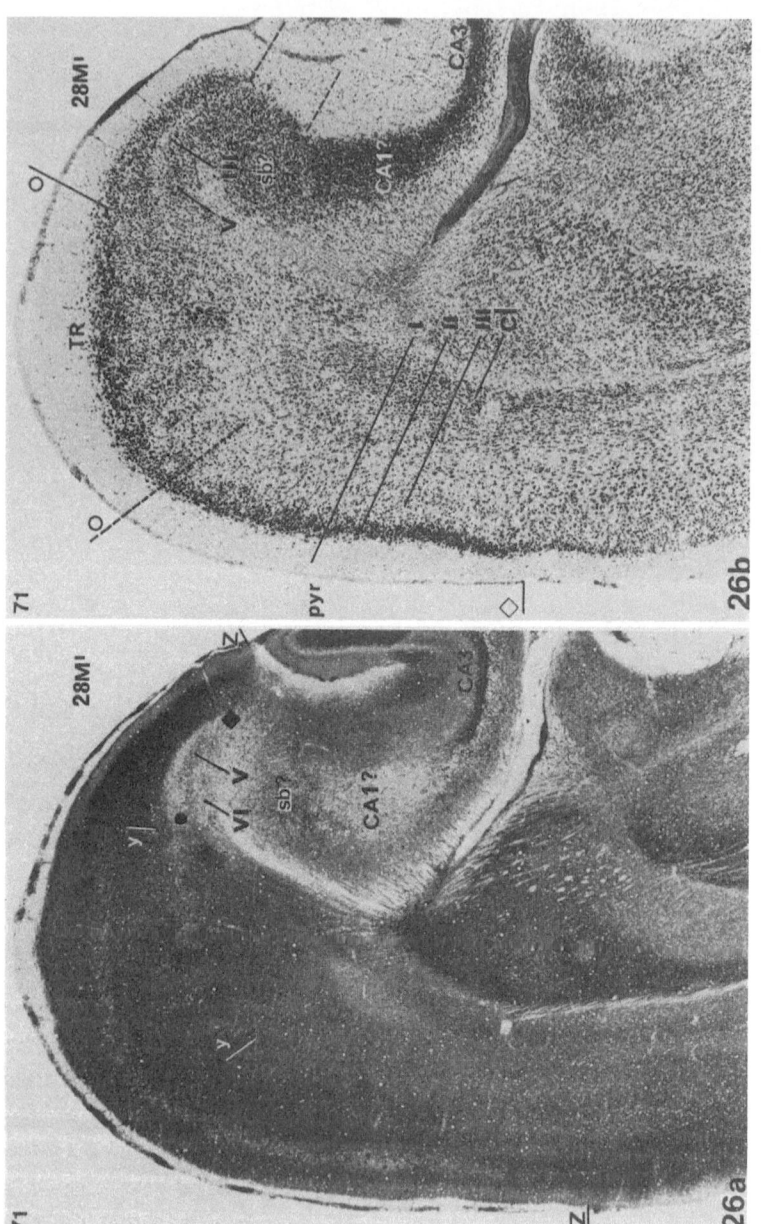

Fig. 26a and b. Horizontal series. × 18

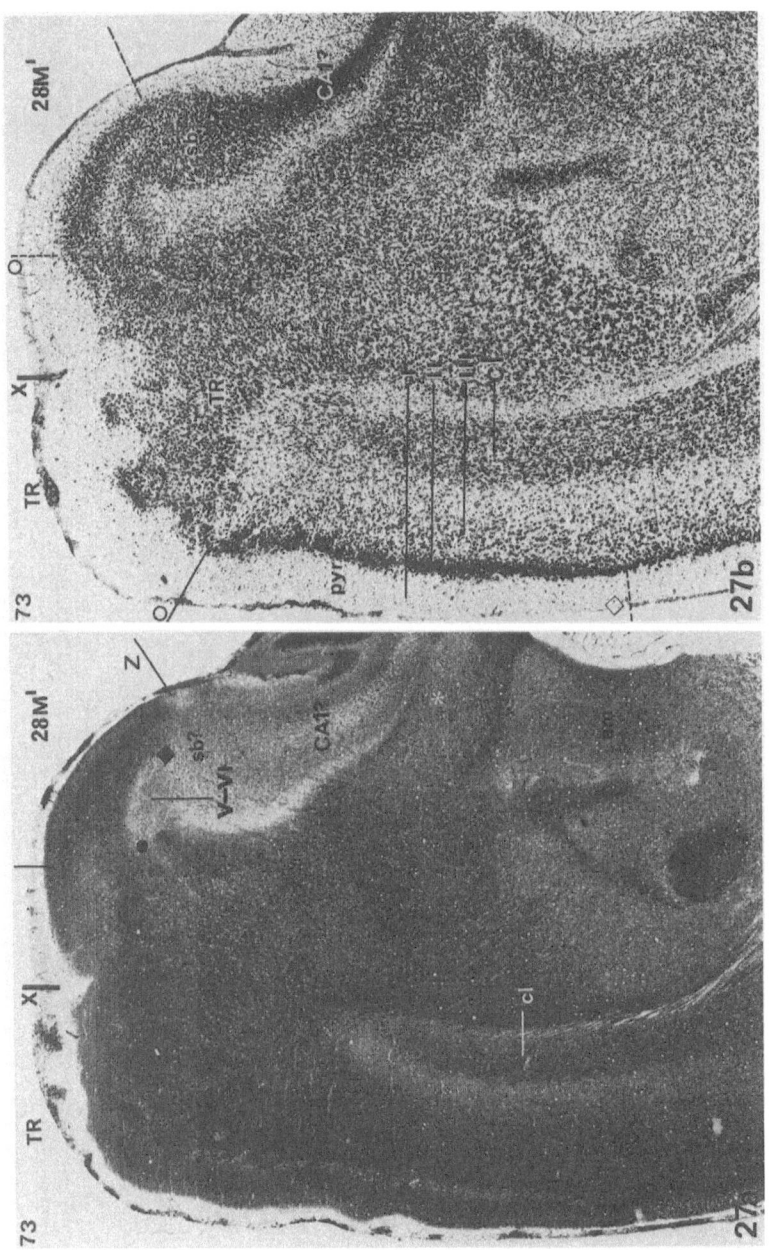

Fig. 27a and b. Horizontal series. × 18

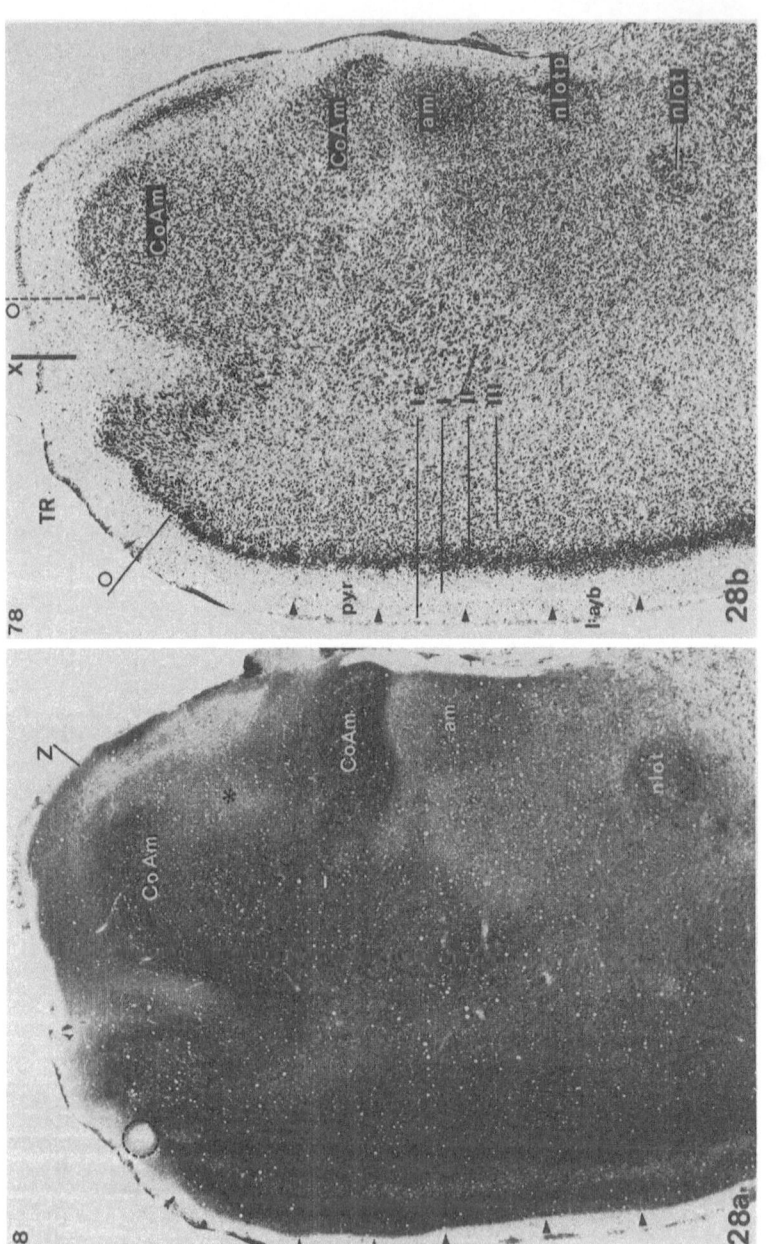

Fig. 28a and b. Horizontal series. Asterisk on posteromedial basal nucleus of the amygdala (Scalia and Winans, 1975), s. area parahippocampalis, Pam Ch (Stephan, 1975, p. 358) × 18

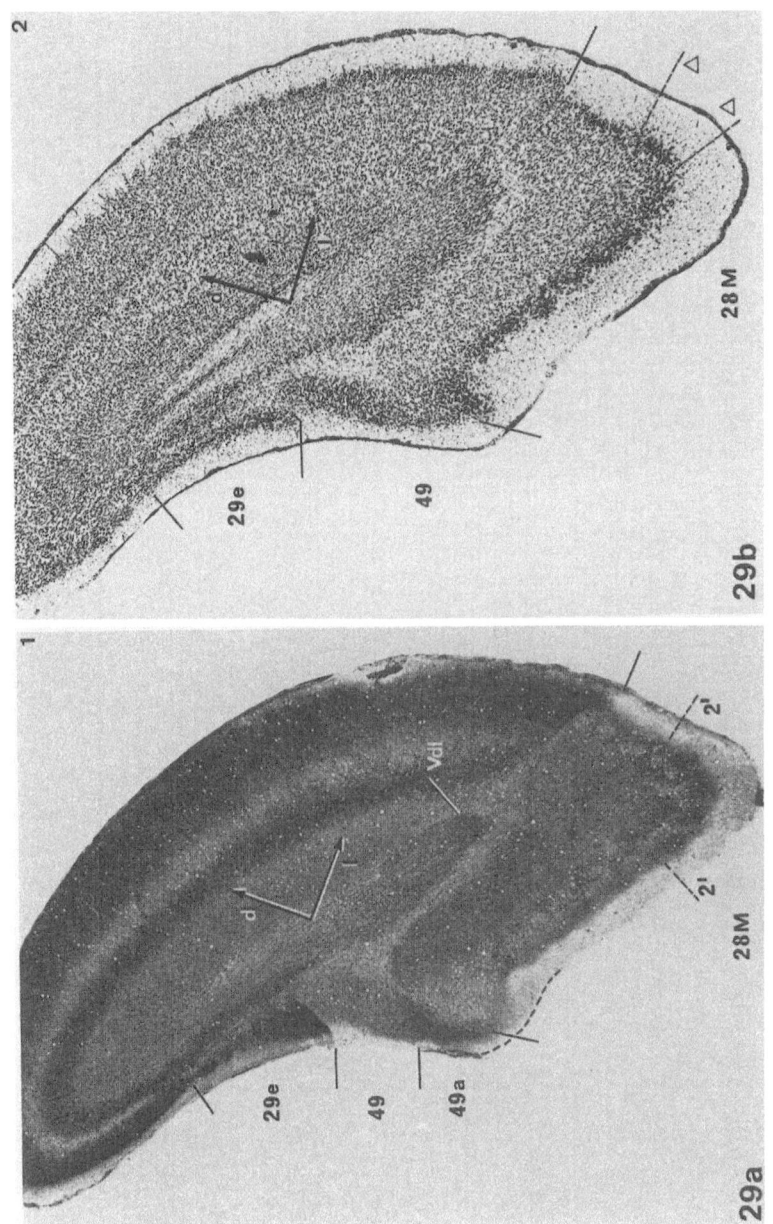

Fig. 29a and b. Frontal series. × 18

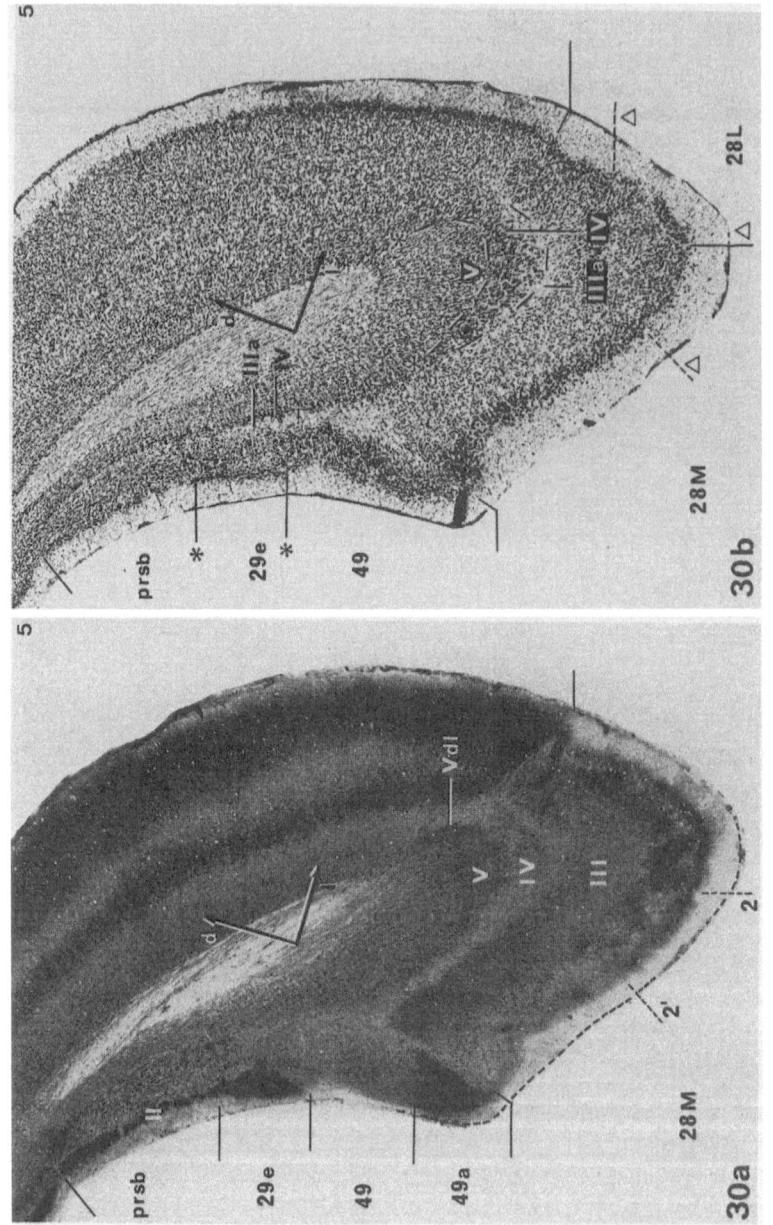

Fig. 30a and b. Frontal series. × 18

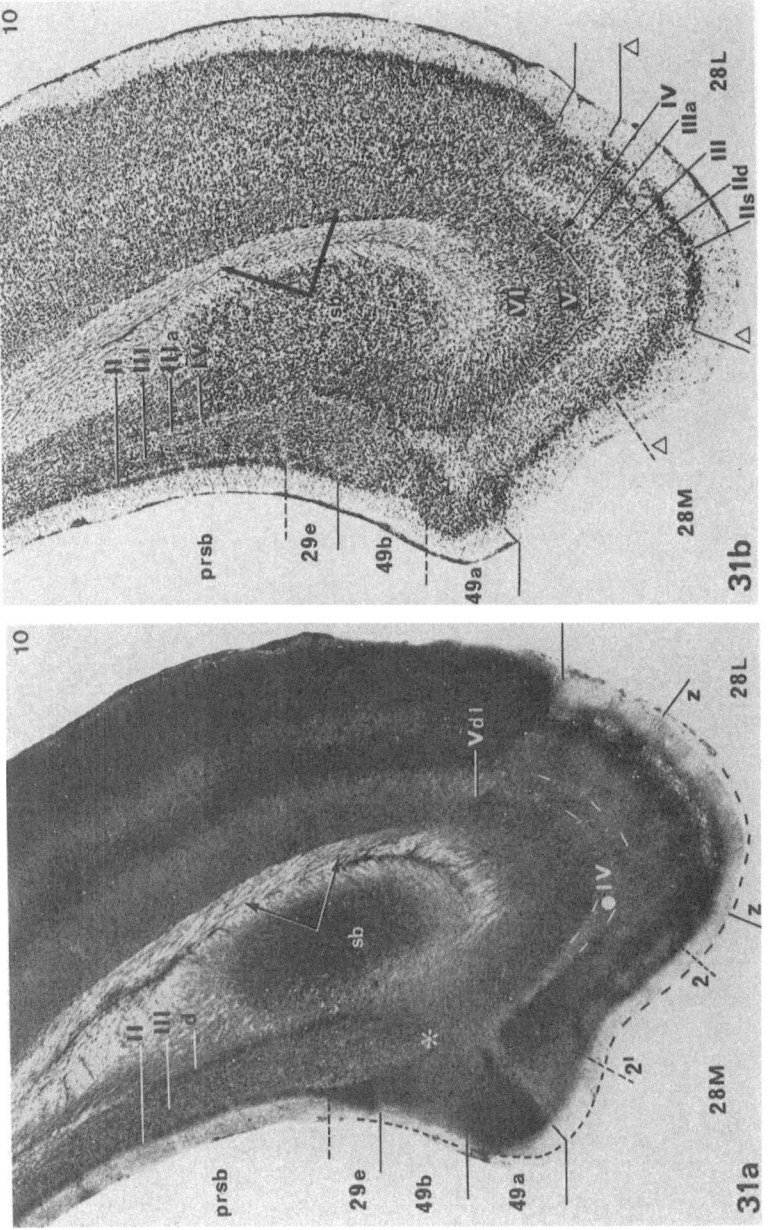

Fig. 31a and b. Frontal series. × 18

61

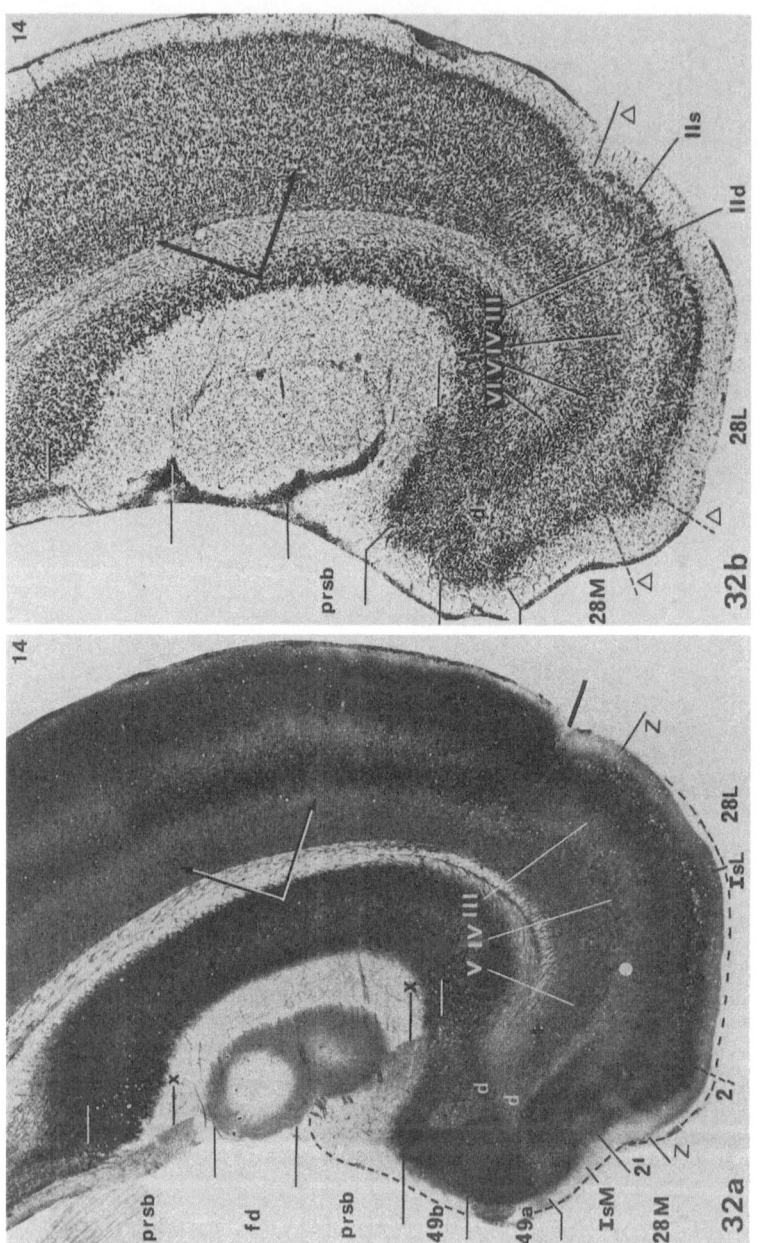

Fig. 32a and b. Frontal series. × 18

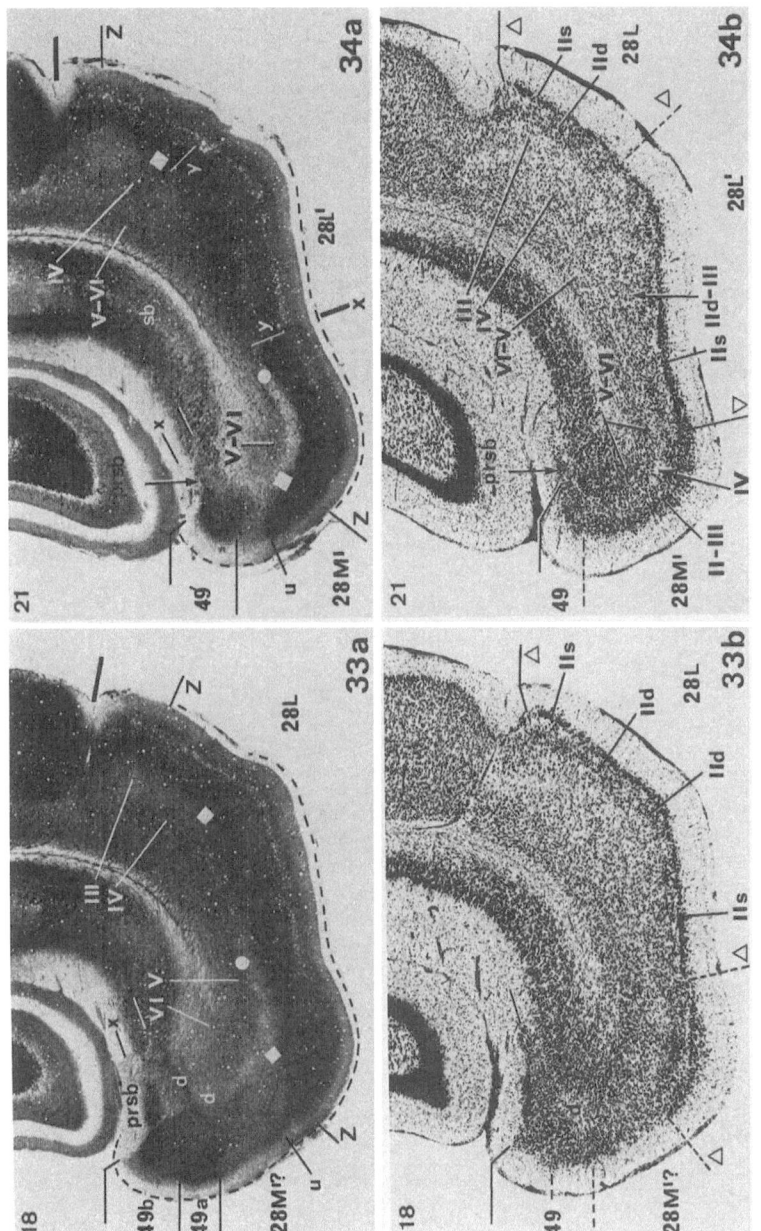

Figs. 33 and 34. Frontal series. × 18

63

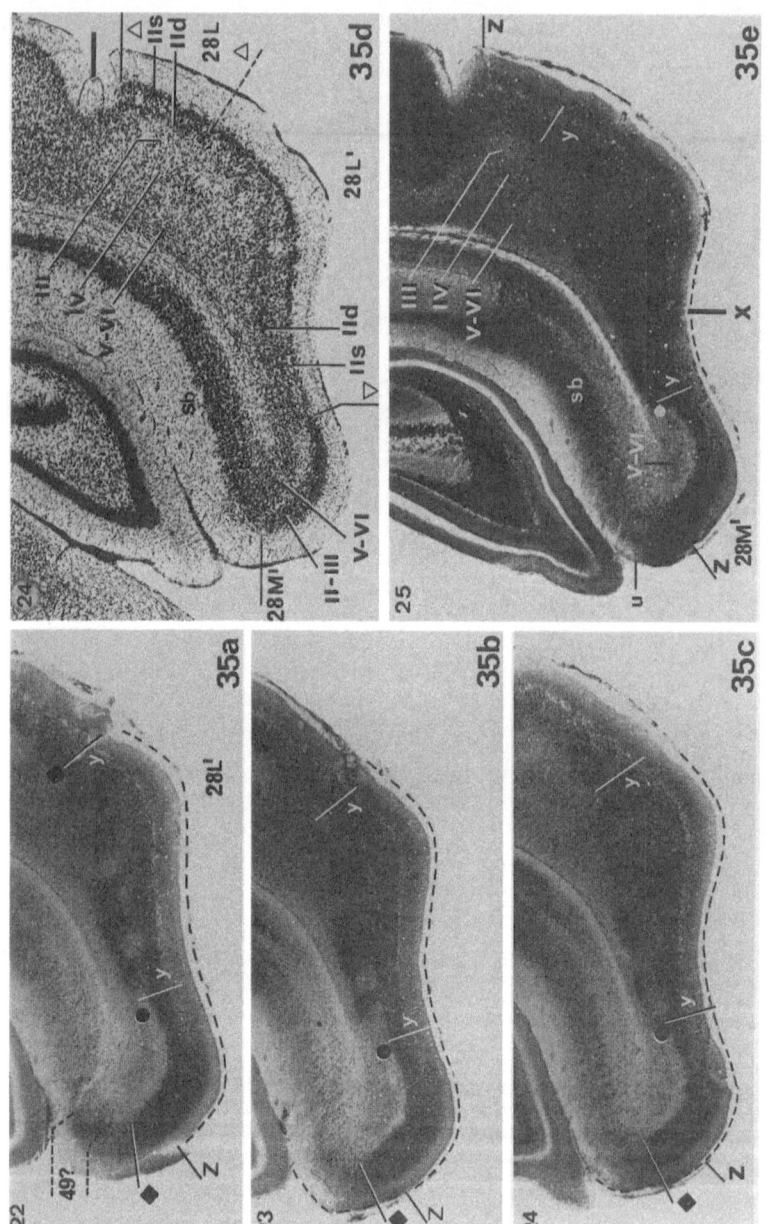

Fig. 35a—e. Frontal series. × 18

64

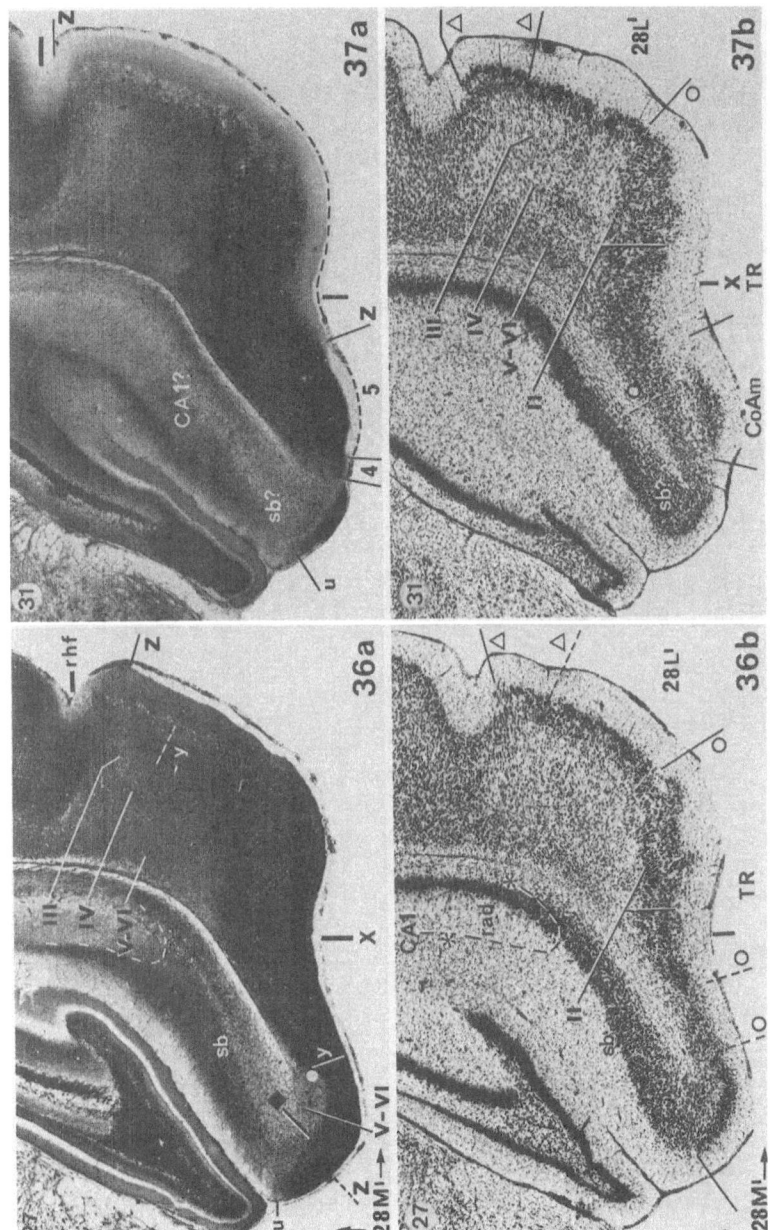

Figs. 36 and 37. Frontal series. × 18

65

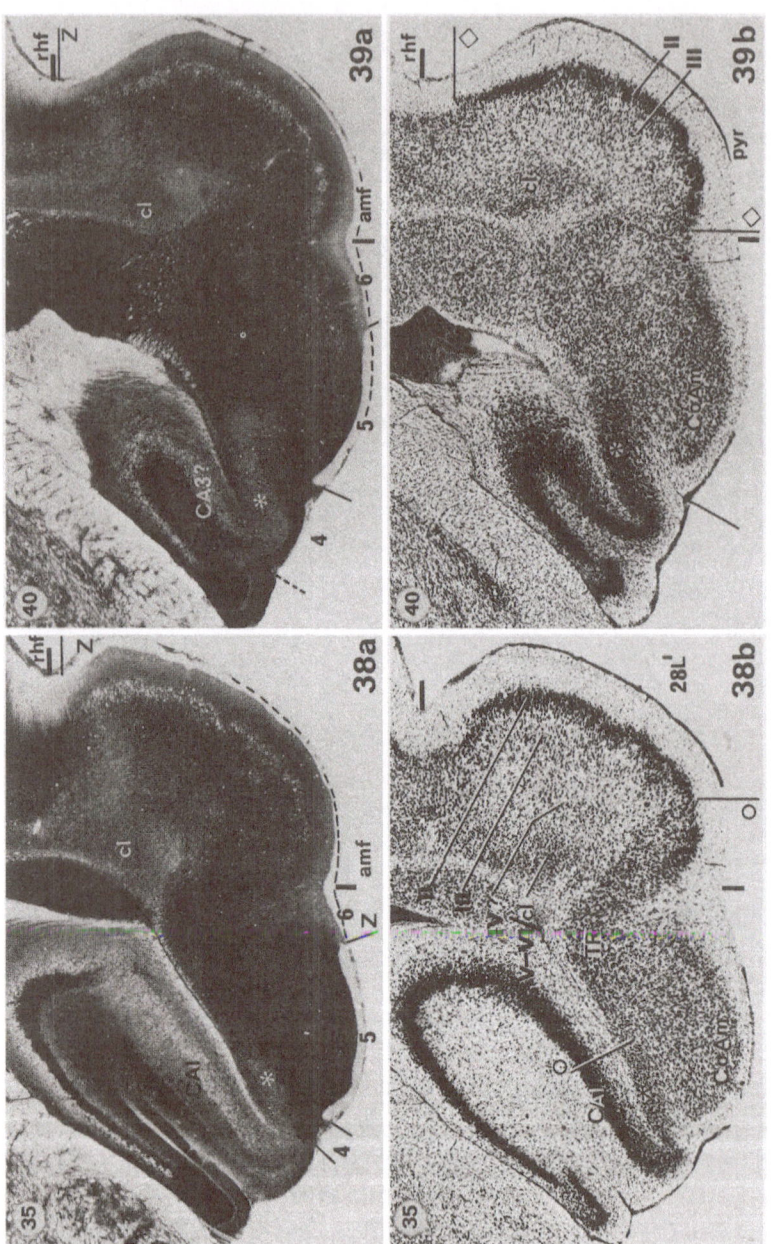

Figs. 38 and 39. Frontal series. Asterisk on posteromedial basal nucleus of the amygdala (Scalia and Winans, 1975), s. area paralippocampalis, Pam Ch (Stephan, p. 358). × 18

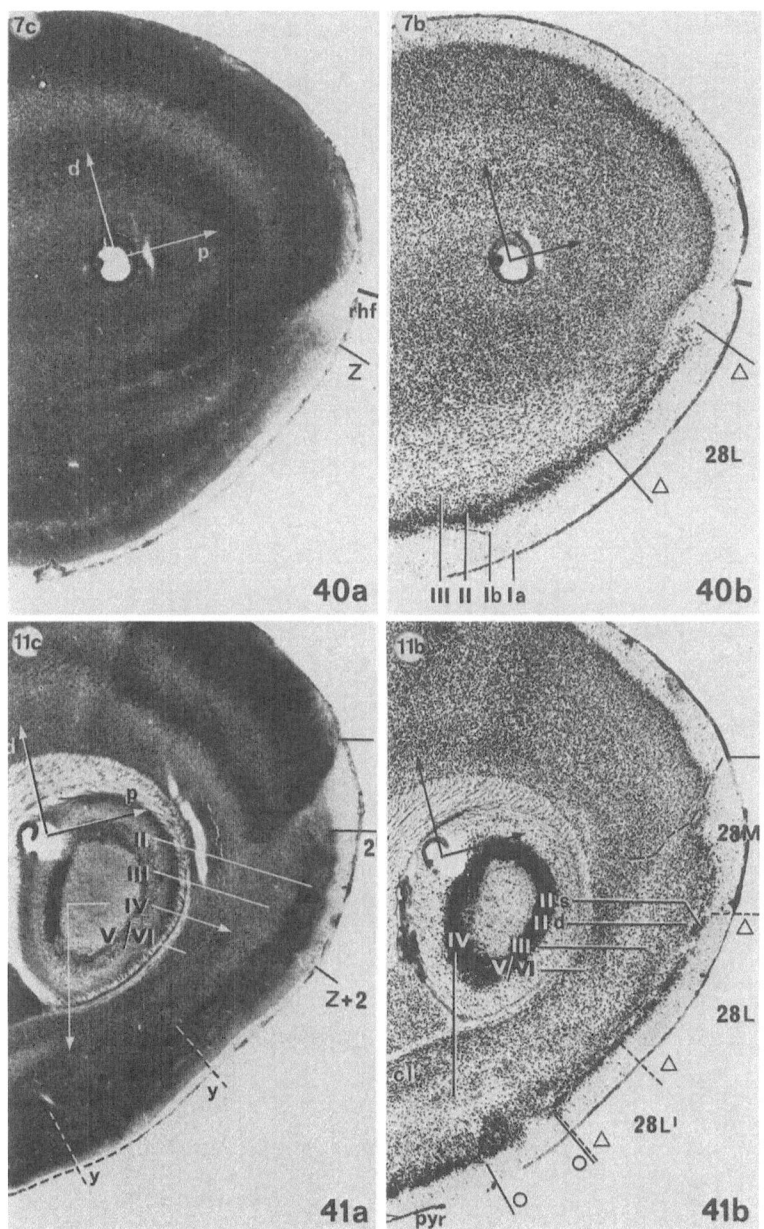

Figs. 40 and 41. Sagittal series. × 11

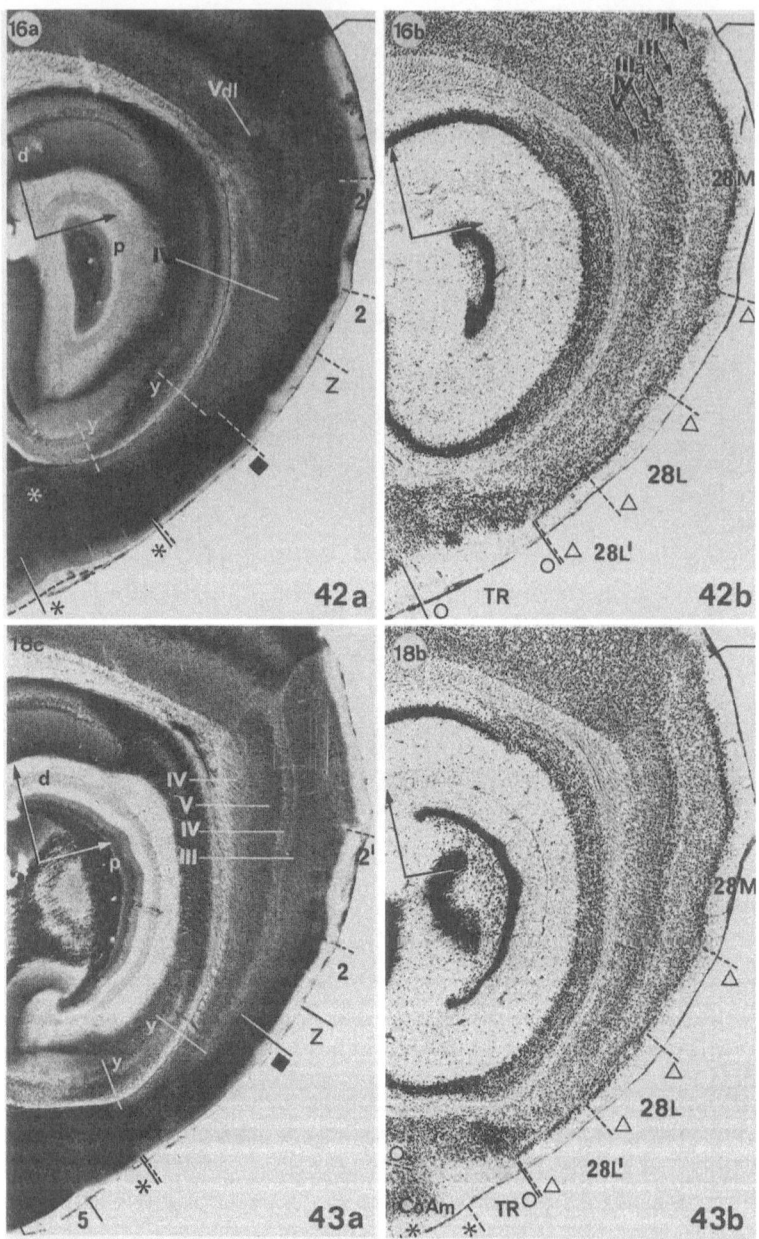

Figs. 42 and 43. Sagittal series. × 11

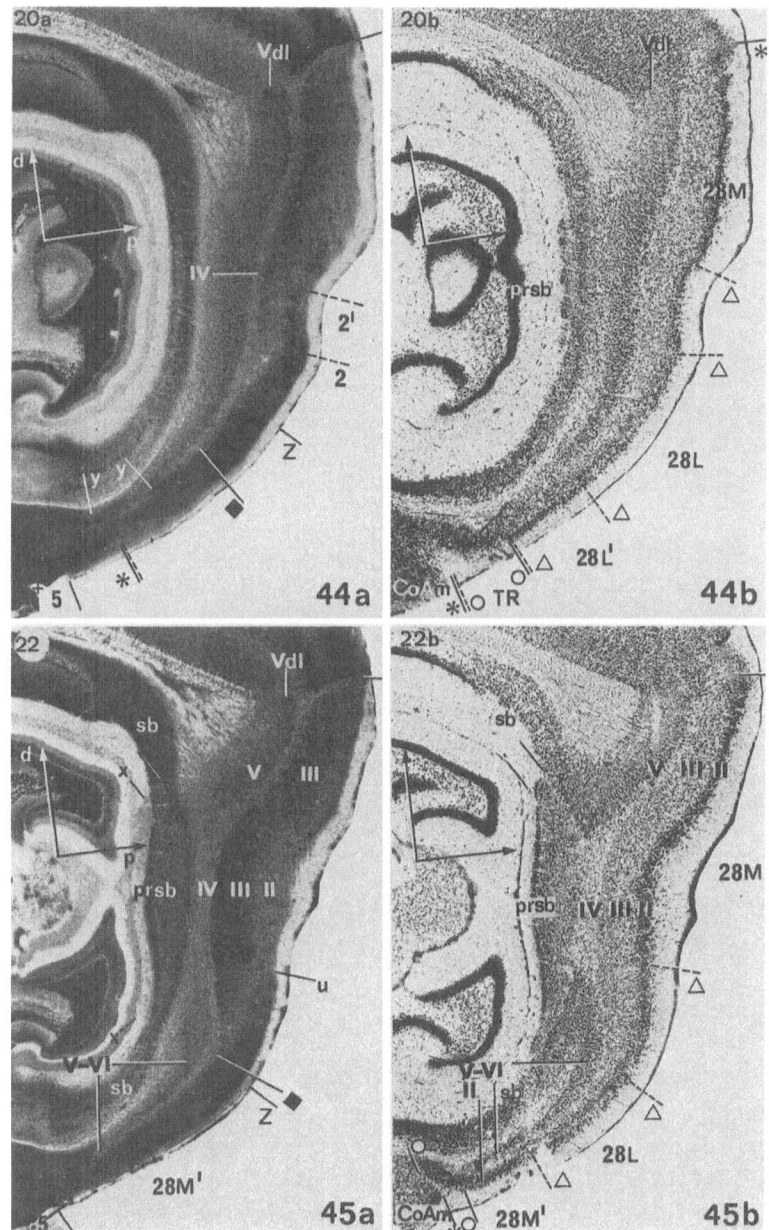

Figs. 44 and 45. Sagittal series. × 11

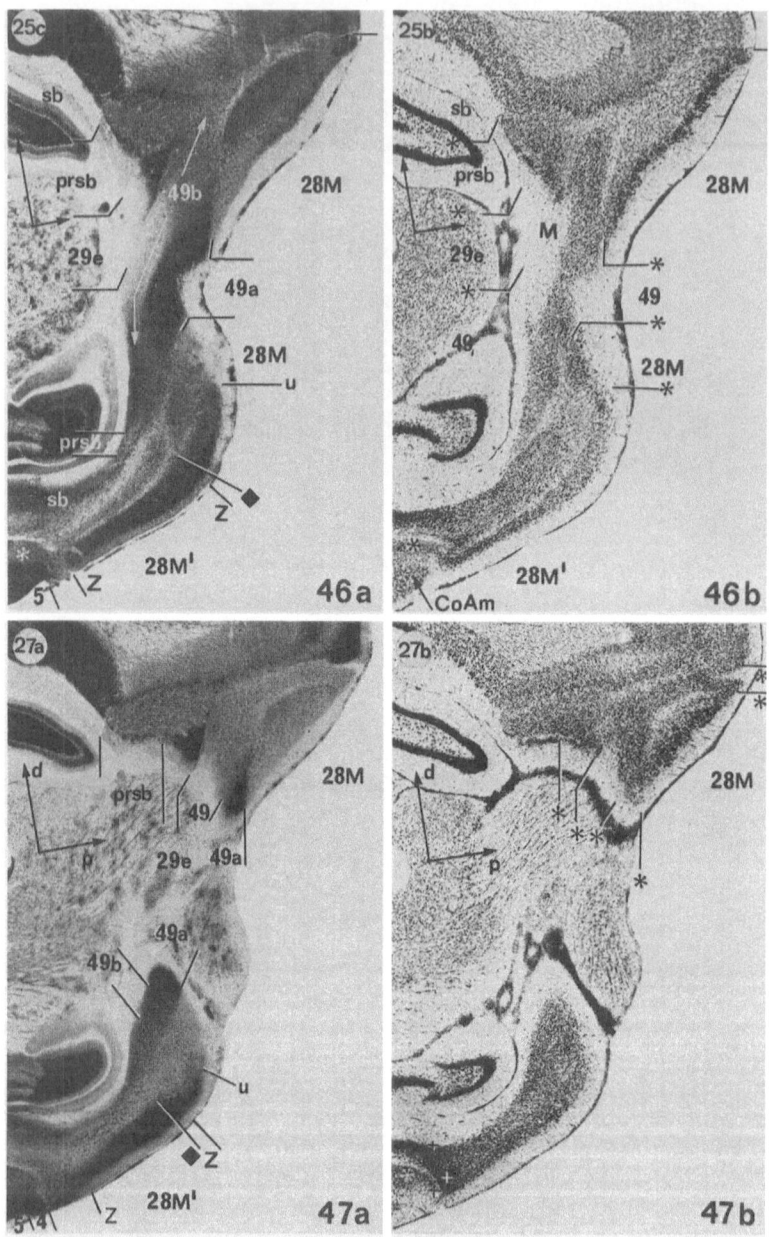

Figs. 46 and 47. Sagittal series. Asterisk on posteromedial basal nucleus of the amygdala (Scalia and Winans, 1975), s. area parahippocampalis, Pam Ch (Stephan, 1975, p. 358). × 11

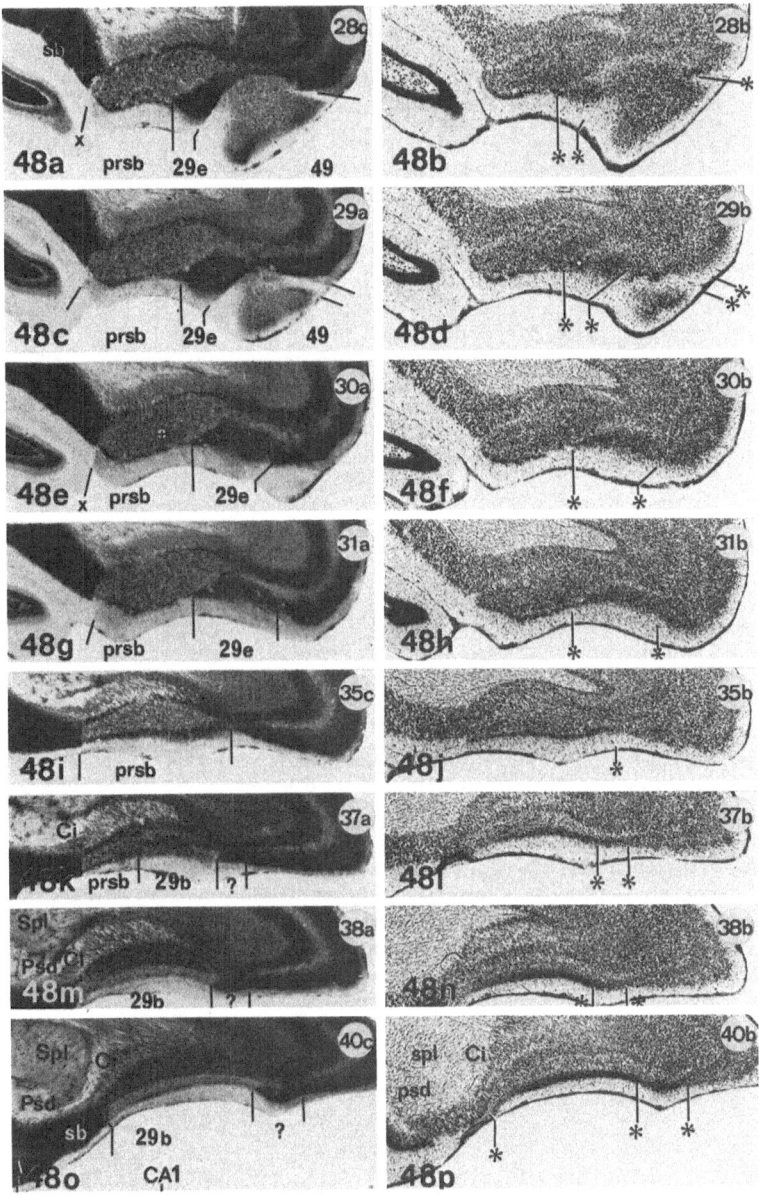

Fig. 48a–p. Sagittal series. × 11

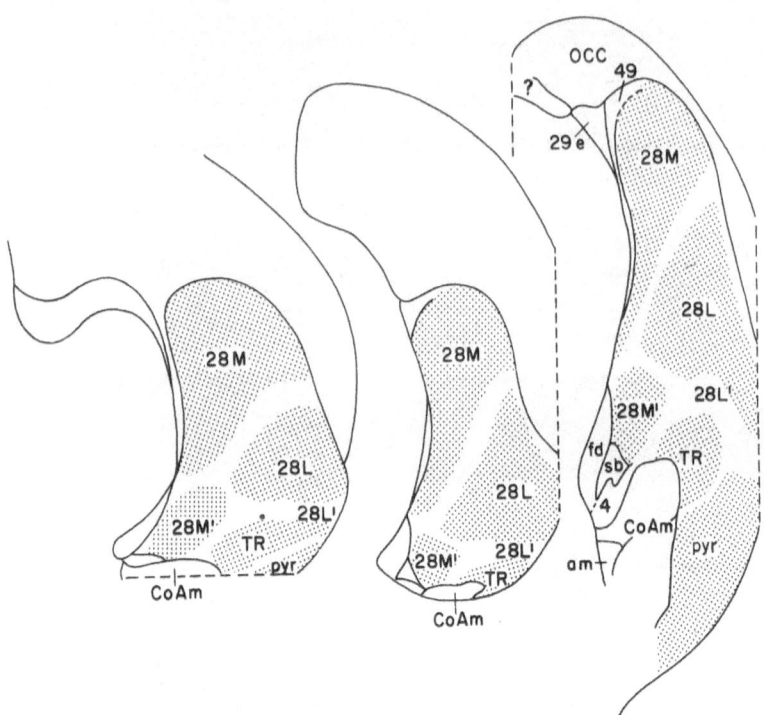

Fig. 49a—c. Diagrams, based mainly on the cytoarchitectonic borders in Figs. 4, 9 and 10, to show with less detail the subdivisions of the entorhinal area and the delimitation *vis a vis* the pyriform cortex which seem justified according to the present study. The diagram largely confirms the previously established subdivision into areas *28M* and *28L*, although allowance is made for a transitional zone between these two areas as well as between the other subareas. The subdivisions *28M'*, *28L'* and *TR* have not been worked out in any detail in the previous literature. Area *TR* is so different from the other entorhinal areas that one may question its inclusion with them. Area *28L'* is the transitional zone between area *28L* and the pyriform cortex (white zone between them on the diagram to the right). Note that the first two diagrams represent posterior aspects of the areas studied while the last diagram represents a posteroventral aspect

Subject Index